Selling Your Idea or Invention

Selling Your Idea or Invention

The Birthplace-to-Marketplace Guide

Sonny Bloch

with

Gary Ahlert

A Birch Lane Press Book

Published by Carol Publishing Group

A Birch Lane Press Book
Published by Carol Publishing Group
Birch Lane Press is a registered trademark of Carol Communications, Inc.
Editorial Offices: 600 Madison Avenue, New York, N.Y. 10022
Sales & Distribution Offices: 120 Enterprise Avenue, Secaucus, N.J. 07094
In Canada: Canadian Manda Group, P.O. Box 920, Station U, Toronto,
   Ontario M8Z 5P9
Queries regarding rights and permissions should be addressed to
Carol Publishing Group, 600 Madison Avenue, New York, N.Y. 10022

Carol Publishing Group books are available at special discounts for bulk purchases, sales promotions, fundraising, or educational purposes. Special editions can be created to specifications. For details contact: Special Sales Department, Carol Publishing Group, 120 Enterprise Avenue, Secaucus, N.J. 07094

Manufactured in the United States of America

10 9 8 7 6 5 4 3 2 1

Library of Congress Cataloging-in-Publication Data

Bloch, H. I. Sonny
   Selling your idea or invention : the birthplace-to-marketplace guide / by Sonny Bloch with Gary Ahlert.
      p. cm.
   "A Birch Lane Press book."
   ISBN 1–55972–244–4
   1. Inventions—Marketing. 2. Patents. 3. Trademarks.
I. Ahlert, Gary. II. Title.
T339.B63   1994
608'.068'8—dc 20   94–19043
                                                            CIP

To Hilda,
My Inspiration

SONNY BLOCH

To J,
To have prevailed in the face of cruelty and deceit,
 is an achievement.
To have succeeded in fulfilling a dream, is justice.

GARY AHLERT

The authors wish to thank Susan Passarelli
and Michael Andrade for their help
in completing this book.

Contents

Introduction   xv

Part I
Intellectual Property Basics

Chapter 1: Owning Ideas   3

The Million-Dollar Idea   3
Ideas as Property   4
Trade Secrets   6

Part II
Patents

Chapter 2: Some Patent Basics   11

Is It Possible?   11
What Need Does It Fill?   11
Getting It Down   12
Is It Cost-Effective?   13
What *Not* to Do Next   13
    Do *Not* Contact Businesses   *13*
    The Nondisclosure Agreement   *14*
    Do *Not* Contact a Local Attorney   *14*
    Do *Not* Attempt to Apply for a Patent Without an
      Attorney or Patent Agent   *14*
What *To* Do Next   15
The Cost of Inventiveness   16
Disclosure Documents   16

Chapter 3: The Patent Process   18

What Is a Patent?   18
- Rights *Not* Granted   *19*
- Right of Exclusion   *19*

United States Patent Laws: A Brief History   20

The U.S. Patent and Trademark Office: Yesterday   21

The U.S. Patent and Trademark Office: Today   21

How the Patent Office Works   22
- The Commissioner   *22*
- Examiners   *22*
- Board of Patent Appeals   *23*
- Additional Services   *23*

What *Can* Be Patented?   23

What *Can't* Be Patented?   24
- Business Methods and Printed Matter   *24*
- Atomic Weaponry   *24*
- Certain Mixtures   *24*
- Ideas   *25*
- Previous Public Use (One-Year Limit)   *25*
- Similarities to Existing Inventions   *25*

Who May Apply for a Patent   26
- Joint Patents   *26*
- Conflicting Inventors   *27*

Chapter 4: The Patent Application   28

Neatness Counts   29

No Money-Back Guarantees   29

The Application Itself   29
- The Oath or Declaration   *29*
- The Specification   *30*
- Title   *30*
- Clear Language   *31*
- Distinguishing the Invention   *31*
- Variation on Existing Inventions   *31*
- Technical Abstract   *31*
- Summary   *31*
- Description of Drawings   *32*
- Claims   *32*
- Multiple Dependent Claims   *33*

Conformity *33*
Drawings (Including Standards for Drawings) *33*
Sheet Identification *37*
Complying With the Rules *38*
Models, Exhibits, and Specimens *38*
Compositions of Matter *38*
Allowance and Issue of Patent   38
Fees   38
National Defense Exception   39
Maintenance Fees   39
Expiration   40
Correction of Granted Patents   40
Patent Marking   41
Other Types of Patents   41
Design Patents *41*
Design Patent Cautions *42*
Plant Patents *42*

Chapter 5: Licensing and Marketing   45

Patent Pending Status   45
Invention Company Rip-Offs   46
The Two Options   47
Agents   47
How Do You Find a Good Agent? *48*
Minimal Fees *48*
Nondisclosure Statement *49*
How Much Do Agents Charge? *49*
Types of Contracts   49
What Is Licensing? *50*
Royalties *50*
Liability *50*
Advances *51*
Before Signing Anything   51
Selling the Invention Outright   52
Profiting From Repressed Ideas   52
Self-Production   53
Self-Manufacture Concerns *53*
Increased Profits *54*
When It May Be Better to Go It Alone *55*
Starting Small   55

Raising Funds   56
    Letter of Intent 57
    Borrowing From Friends and Family 57
    Patent Attorney as Temporary Partner 57
    Raising Start-Up Funds/Going It Alone 58
    Beware of Scams Again 58
    Other Fund-Raising Ideas 58
Advertising Yourself and Your Product   59
Selling Your Patent   60
    Recording of Assignments 61
Joint Ownership   62
Granting License   62
Patent Infringement   62
Conclusion   64
    An Inventive Tale 64
Patent Appendix I: Publications of the Patent and
    Trademark Office   66
Patent Appendix II: Correspondence With the
    Trademark Office   69
Patent Appendix III: The Search Room and
    Depository Libraries   71
Patent Appendix IV: Patent Fee Schedule   76
Patent Appendix V: Foreign Patent Protection   81
Patent Appendix VI: A Sample Patent   86
Patent Appendix VII: Patent Application Forms   90
Patent Appendix VIII: A Sample Agent Agreement   94
Patent Appendix IX: A Sample Licensing Agreement   99
Patent Appendix X: A Sample Nondisclosure
    Agreement   105

Part III
Copyrights

Chapter 6: Copyright Basics   111

What Is a Copyright?   111
History   112
The Five Rights   113
What *Can* Be Copyrighted?   114

Sufficient Creative Effort   115
   A Fixed and Tangible Expression   *115*
   Original Works of Authorship   *115*
What *Can't* Be Protected?   116
The Public Domain   117
Fact and Fiction   118
Who *Can* Claim Copyright?   118
   Collaborations   *118*
   Collective Works   *118*
   Works for Hire   *119*
Minors   120
Limits on Copyright Protection   120
Fair Use   121

Chapter 7: Copyright Notice   124

Different Forms for Different Media   125
Form of Notice for Visually Perceptible Copies   125
Form of Notice for Phonorecords of Sound Recordings   125
Position of Notice   126
Works by the U.S. Government   126
Unpublished Works   127
If the Notice Is Omitted or Incorrect, 1978–1989   127

Chapter 8: Copyright Registration   129

When Works Can Be Registered   129
Renewals   130
Registration Procedures   130
Deposit Requirements   131
Special Deposit Requirements   131
Unpublished Collections   132
Filing Corrections   133
Mandatory Deposit for Works Published in the
   United States   133
Who May File an Application Form   134
Mailing Instructions   134
What Happens if the Three Elements Are Not
   Received Together   135
Fees   135
Effective Date of Registration   135
What to Expect From the Copyright Office   136

Search of Copyright Office Records   136
How Long Copyright Protection Endures   137
    Works Originally Copyrighted on or After
      January 1, 1978   *137*
    Works Copyrighted Before January 1, 1978   *137*

Chapter 9: Publication and Licensing   139

What Is Publication?   139
How to Proceed With Your Copyrighted Work   140
Is an Agent Necessary?   141
Getting an Agent Anyway   142
Producing an Attractive Package for Your Work   143
Types of Agreements   144
Transfer of Copyright   144
What Can Be Transferred   145
Types of Transfer Contracts   145
    Exclusive Licenses   *145*
    Nonexclusive Licenses   *146*
Contract Points   146
Compulsory Licensing   148
Termination of Transfers—The Thirty-five-Year Limit   149
Recording Transfer Contracts   149
Infringement   149
The Future of Copyright Law   151
Copyright Appendix I: A List of Copyright Forms   153
Copyright Appendix II: Copyright Forms   155
Copyright Appendix III: International Copyright
    Protection   185
Copyright Appendix IV: Library of Congress Catalog
    Numbers   188

Part IV
Trademarks

Chapter 10: Trademark Basics   191

Trademark Rights   191
When Would I Need a Trademark?   192
Trademarking a Product Name for a Patent   192
Trademarking a Character Name   193

The Source of Trademark Protection   193
Benefits of Registration   193
The Registration Process   194
  Initial Determination   *195*
  Approval   *195*
  Intent To Use   *196*

Chapter 11: Trademark Applications Up Close   197

The Written Application Form   197
  The Heading   *198*
  Applicant   *198*
  Identification of Goods or Services   *198*
  Basis for Application   *198*
  Execution   *199*
The Drawing   200
  Drawing Heading   *200*
  Typed Drawing   *201*
  Special-Form Drawing   *201*
  Size of Special Drawings   *201*
  Use of Color in Special Drawings   *202*
Specimens (Examples of use)   202
  Nature of the Specimens   *202*
The Filing Fee   203
Further Requirements for Intent-to-Use Applicants   203
Extensions   204
Grounds for Refusal of the Application   204
The Supplemental Register   205
Trademark Search Library   205
Who May File an Application   206
Notice   206
Renewal and Maintenance Terms   206
Conclusion   207
Trademark Appendix I: Trademark Application Forms   208
Trademark Appendix II: Foreign Applicants and
  General Information   230
Trademark Appendix III: The International Schedule
  of Classes of Goods and Services   231

Index   235

"Imagination is more important than knowledge"
—ALBERT EINSTEIN

Introduction

The purpose of this book is to provide you with detailed yet simple information on how to obtain copyrights, patents, and trademarks, and also how to get your invention or creative work to the marketplace. It is my hope to provide all the answers to not only the questions you *might* ask, but also those you *should* ask. Step by step, this book will guide you through all of the procedures necessary to develop and protect your idea, as well as provide a comprehensive overview of *intellectual property law*. This information has been culled from a variety of sources, including many successful inventors and other creators, patent attorneys, and—of course—the United States Patent and Trademark Office.

It is my sincere hope that this book will enlighten you about your rights, as well as inform you of your options. There are many dangers and pitfalls awaiting you out there: Let us presume that forewarned is forearmed. Yet, even more than safeguarding the would-be inventor from possible mistakes, this book should help you to get your idea or invention out of your head and into the marketplace.

You see, many of us have ideas, but very few of us act on them. There is, of course, a litany of excuses for inaction—such as thinking "It will never work," or deciding that the idea we thought was wonderful just a minute ago now seems just plain dumb. So please remember: It's always important not to demean your own ideas or think your concepts too small or trivial to pursue. The inventor of the paper clip, a cheap little product that has generated many millions of dollars, sold all his rights to it for $50—to pay off a debt.

We live in an ever more capitalist world, in the early days of the Age of Information. Society is driven by ideas, and each has its place and its merit. So, if your idea is for a new electric car that will help save the environment—well, fantastic! But even if it's "only" for a small novelty or toy that can bring a lessened burden or increased enjoyment to someone's life, *that* is great, too! Keep in mind that along with the pleasure of creating something new, or even of improving something old, there also can come the possibility of great financial reward.

PART I

Intellectual Property Basics

CHAPTER 1

Owning Ideas

Everyone has them at one time or another. Some are referred to as being "million-dollar," others simply as "good." Some lead to greatness, others to obscurity. I am talking, of course, about those ephemeral things called *ideas*.

Where do ideas come from? Perhaps from the right combination of chance, need, and an initially subconscious stirring of that transitory virtue called creativity. It may be that when we think about a certain problem or puzzlement for long periods, sometimes even dreaming about it at night, our mind naturally generates solutions via intuitive perceptions concerning what we're burdening it with. Remember that I said "It may be"—since without the *right* combinations of circumstances, all our mental wonderments can prove meaningless. Ah—but when a combination *does* click: *voilà!*

There are various exercises you can perform in order to get yourself thinking more creatively. (In fact, an entire section of most any bookstore will be devoted to the subject.) In the end, however, you're just as likely to have a brainstorm on an elevator as behind a desk. So, where *do* creative ideas come from? The real answer is: Nobody knows! The most important thing—once you *do* have such an idea—is recognizing its possibilities. The second most important is knowing what to do next.

The Million Dollar Idea

So one day you're driving in your car, or sitting down to a meal, or enjoying a nice hot shower, when suddenly, seemingly out of the blue, you're struck by a one-in-a-billion idea (a stunning concept for a brand-new gizmo or a new and improved whatchamacallit)—or maybe the missing piece of that pot-boiler detective novel you've always wanted to write suddenly starts dictating itself into your mind's ear. As you sit or stand there, brow furrowed, brain pulsating, bouncing that thing around in your head, you become absolutely convinced that it will improve the lot of mankind by making some terrible task easier. Or simply bring a smile of pleasure into someone's day.

But what do you do *next?* Where do you *go* with it? Whom do you *show* it to? How do you *make money* with it? How do you *protect your rights* to it? And, maybe even more important: What *shouldn't* you do? What mistakes might you make that would *hurt* the chances of bringing your idea to fruition, or (maybe far worse) put the profits from it into someone else's hands?

Well, the first step is obvious enough: You should determine exactly what *sort* of idea it is you've had. and under what category it can be *protected*.

Ideas as Property

If you were a shoemaker, your products—shoes—would remain in your possession until you chose to sell them. The fact that they were in your possession would enable you to protect them by locking your shoe shop at night. A variety of laws recognize your right to your property, and help you protect your ownership. But what about an idea? Ideas have no physical substance—you *can't* lock them up at night—and so it wouldn't be very wise to sell them "one to a customer," since the value of any given idea could range from nothing to millions of dollars.

Despite the apparent insubstantiality of ideas, we recognize them both legally and morally, as the "property" of their creators, or owners. Ideas, therefore, are referred to as *intellectual property*.

Intellectual property is a fancy legal name that gives ideas, concepts, and inventions much the same sort of protection as physical property. It doesn't matter what kind of idea it is, or what use it is intended for. It can be for *your* benefit, for the benefit of *mankind,* or simply for *fun.* Just about anything you can create becomes intellectual property—be it a song, a book, a new pencil sharpener, or a Pet Rock.

Naturally, since ideas aren't the same as physical property, the laws that pertain to their protection are different as well. *All* ideas are both covered and protected by *intellectual property rights and laws.* When used with intelligence and awareness, these help to prevent people from stealing your ideas and also from using them without your permission. If someone *infringes on* an idea you rightfully own, you have the right to go up against that party in court and sue for damages.

Intellectual property rights basically fall into four areas:

Patents: This covers inventions of all sorts. It would be impossible to list all the types of inventions, but a general list might well consist of such things as:

- a new type of computer chip
- processes
- machines
- articles of manufacture
- compositions of matter

Trademarks: This covers the rather specific area of logos and logo designs. A trademark consists of a name or artwork, or combination of the two, used as an identifying mark (or *logo*) that gives a particular identification to your business or product. In other words, a mark that you use for trade purposes.

Copyrights: This covers all types of creative works. Creative works consist of everything from music to film to short stories. Even certain aspects of journalism and histories are included in this category. All these are basically artistic expressions of one form or another. A brief list of creative works includes:

plays, books, magazines, newsletters, bulletins, personal and business correspondence, speeches, scripts, poetry, research reports, computer programs, product packaging, promotional matter, advertising copy, motion pictures, sculpture, photographs, audiovisual programs, art, cartoon strips, toys, scientific and technical drawings, architect's plans, maps, sounds, music

Trade Secrets: This covers business practices. Trade secrets consist of confidential business information. A brief list of trade secrets would include such entries as:

the formula for Coca-Cola, mailing lists, business operations, special techniques used in business functions, and data bases

People commonly confuse patents, copyrights, and trademarks. Although there is *some* resemblance in the rights of these three kinds of intellectual properties, and in the procedures involved in securing those rights, they all serve different purposes. For that reason, separate sections of this book (called Parts) are devoted to exploring each in depth.

If you have an idea for an invention, at this point you might wish to turn to Part II, which covers patents and the patent process.

If however you are concerned in the main with protecting a creative work, you would do better to turn instead to Part III, which discusses copyrights.

Regardless of whether your idea is for an invention or a creative work, you should read Part IV, which is on trademarks—since the information presented there applies to copyrights and patents instances.

Trade Secrets

Since the category of trade secrets is the simplest and most straightforward of the intellectual property categories, we will briefly cover it now.

The notion of the trade secret is based on the legal principle that you cannot steal the tools a person uses to make his or her living. Exactly what constitutes a "tool" is, of course, open to some debate—but a general definition used in the federal courts and in most states is as follows:

A trade secret is any formula, pattern, device or compilation of information which is used in a person's business, and gives [him or her] an opportunity to obtain an advantage over competitors who do not know or use it.

A trade secret, then, could be your client lists, a particular method for marketing information or even secret technology. Silly Putty, for example, has never been patented, but the formula for *making* it falls under the protection of the trade secret laws.

Definitions aside, a trade secret is the simplest type of intellectual property to protect. Trade secret status is not something you have to file for, apply for, or enlist the help of an attorney to register. You *own* a trade secret simply by virtue of being its originator and by using it regularly in your business! Trade secrets are *considered* property rights. They may be sold, licensed, or transferred by legal contract.

The simplicity of achieving trade secret status by no means implies that the protection of a trade secret is not a serious business. When trade secret cases go to court, the judge's decision generally depends upon your prior efforts to identify the information *as* secret, and on taking appropriate action to protect it. For a one-person operation, this might be as simple as locking (or "padlocking") the filing cabinets that contain your customer lists. For larger businesses, it would involve identifying the areas of your business that use your trade secrets, then asking employees involved in those areas to sign what's known as a *nondisclosure agreement*. The nondisclosure agreement would delineate the trade secrets and forbid your employees from using them—other than in the course of performing their jobs for your business. Nondisclosure agreements typically have *noncompetition* clauses that specify the business areas in which employees would be strictly forbidden to use the secret information.

Naturally, the larger your business is and the more secret processes you use, the more complicated the situation becomes. In some instances, trade secrets created by employees belong to the business they were created for. This arrangement is referred to as *work-for-hire*, in which case, if an employee is hired to generate ideas, there is a clear agreement that such ideas become the property of the employer. In such instances, the employer is considered by law to be the inventor, and the employee loses all rights to the invention. (In other instances, however, the secrets created do *not* automatically belong to the employer. This depends on the type of secret created, and the defined duties of the employee.)

While this is a fascinating area of intellectual property law that can reach into the realm of industrial espionage, the subtleties involved are beyond the scope of this general guide. The important thing to realize is that the data you cull, and the processes you create for doing business, belong to *you*—unless you agree otherwise.

PART  II

Patents

CHAPTER 2

Some Patent Basics

In this section we explore the various steps involved in protecting your invention: seeking help, the patent process, and marketing and licensing your invention. But before we proceed through the lengthy, somewhat expensive and complex patent process, there are a few questions you really should ask yourself about your idea.

Is It Possible?

The first question is whether or not your idea falls within the realm of the possible. A matter–antimatter warp drive is a wonderful notion, but it remains to be seen whether or not it would work beyond the world of science fiction television shows.

Some people have brilliant ideas for things they don't know how to build. For example, if your idea is for a device that will electronically locate missing socks, but you have no idea how the electronics would work or how you would go about building such a unit, you would either have to learn electronics or acquire a partner to help you make the gizmo work—and, after all that, it might turn out that an electronic sock locator simply isn't feasible.

Unfortunately, ideas are not always practical. You really have to think about that when thinking them out.

What Need Does It Fill?

A related and equally important question is: What need does your invention fill? Most people get their best ideas by trying to fulfill a real need they encounter in their own lives. Often, at work or at home, people ask themselves, "How can I do this better?" Or how can I improve the device I'm already using? An invention based on this sort of experience has a built-in usefulness, since it was created with a particular need in mind. The next question to ask, then, is, "How many other people have the same need?" (Will this idea be useful to others, or something that works just for me?)

It is entirely possible, of course, that your invention might *not* involve something that already exists. It could well be an idea for something entirely new—something the world actually has never seen before. In that case, you might ask yourself whether anyone will even want to see it. If the answer is yes, your invention may be able to create its *own* need. Deodorants, for example, weren't considered a universal requirement until after they'd become available in widely marketed forms.

Notice that while the idea itself must be practical, the need it fills *doesn't* have to be. Human needs can often be totally *impractical*! For example, your idea could fill a need for entertainment—for pure fun, which (compared to pure work) *is* impractical.

Getting It Down

Once you're confident that your idea is possible to realize, and that it will either fill or create some need, the next step is to knuckle down and try to make it work. Similarly, if it can be made at all, you should carefully draw out plans for your invention—or, better yet, build a working prototype. (The device could even be made out of cardboard or papier-mâché—it doesn't matter, just as long it shows that your idea *will* work.)

In some instances a drawing or prototype might in fact be something short of practical, or perhaps too expensive to make. If diamonds or gold were required in the production, for example, you might have a hard time scrounging some up for your working model. Even so, you'll have to find *some* way to see if your idea will work in the real world. (Remember Thomas Edison's failures? Probably not. And you'll forget yours, too.)

Is It Cost-Effective?

You may discover that a solid platinum strip of metal shaped like a spring will help make superb baked potatoes—but the retail price could be prohibitive to the general consumer, and this factor would dreadfully soon put a severe limitation on your potential market. Ask yourself from the start if the price of building one unit of your invention is less than what someone would be willing to pay for it in its production/run form.

What *Not* To Do Next

Now you've got your working plans (if it is not yet practical or cost-effective to build a model) or model, and you're confident both that your invention fills a need and that its production will be cost-effective. Before we get into the details of what you should do next, let's take a cautionary moment to note a few equally important points about what you should *not* do. It is exquisitely easy to lose the rights to your idea—and what seems to be a perfectly logical, simple next step may turn out to be a disaster waiting to happen.

Do Not *Contact Businesses*

Many people think that the rational thing to do with their idea would be to bring it to the attention of a prominent company already involved in the particular industry that the invention affects. If your idea is for a new paper clip, for example, you might think it wise to call a stationery company, on the assumption that the company will be so thrilled with your one-in-a-million idea that you'll be able to sell it to them for a quick million, sight unseen.

Let's say you take your working model of this new paper clip to ABC Stationery, the largest (fictional) stationery supply company in the country. Well, lo and behold, the first thing ABC Stationery says is, "Don't get excited, Charlie—we get like a zillion of these here every day—and chances are we're probably not interested in yours." But even worse than abject failure might be success! ABC Stationery could say, instead, "Send us a description in writing, or a picture of your new paper clip." And that's where your troubles could really begin.

There are two possible avenues that such a company can scoot down with unprotected ideas:

1. It can say, "Send us the product." You'll send it, and they'll send it back—with a kiss-off letter saying, "Thank you, Mr./Mrs./Ms. Inventor, but, amazingly, we already happen to be working on the exact same idea as you."
2. It will ask you to sign a *nondisclosure agreement.* And often that agreement will take away all your rights!

The Nondisclosure Agreement

That's a fact: A nondisclosure notice says, in effect, that you no longer retain *any* rights to your idea, ever. Whether the business clairvoyantly claims to have been thinking about it, or even if they manage to produce precisely it on their own at some time in the future, you have no rights or protection regarding even your own interpretation of it.

This really leaves you in a Catch-22 situation: You can't sell your idea unless you show it, but if you do show it, you're opening yourself up to the possibility that someone is going to steal it. Things aren't all as bad as they sound—but be aware that great care *is* necessary.

Do Not Contact a Local Attorney

Many people immediately recognize that indeed they *should* proceed carefully in order to protect their rights. Some will, naturally, hire a local lawyer for assistance. Unfortunately, most general-practice attorneys know very, very little about intellectual property laws, that complex, specialized field of jurisprudence devoted to safeguarding ideas. Thus a general-practice attorney with even the best of intentions more than likely would charge you for the wrong advice.

Do Not Attempt to Apply for a Patent Without an Attorney or Patent Agent

Applying for a patent on your own is exceptionally risky. (By the end of this chapter you'll realize exactly how difficult the process is, especially if you try to go it alone.) It's an extremely complex application procedure, subject to dozens of specific rules and spelled out in a language all its own. Very detailed drawings are required, and all the wording must be extremely precise. Aside from all that, the filing fees are quite high, and the government *won't* issue refunds if you've made a mistake. So, unless you have personal expertise in the area, you'll *have* to hire help.

What *To* Do Next

Fortunately, there are three types of professionals who specialize in intellectual property laws. The first is the *patent attorney*. The second is the *patent agent*. The third is businesses that specialize in related fields, such as *licensing firms*. Your next step, then, would be to seek help from one of these sources.

The next logical question is: How does someone find a *good* patent attorney or agent? There is no easy answer. Many patent attorneys and patent agents are fine, moral, upstanding citizens who will do their level best on your behalf. Others may try to cheat you, overcharge you, or ignore you. The best way to find a good patent attorney or agent for the best price is to do the same thing you would do as a careful consumer: Shop around.

For starters, we recommend that you look in your phone book under headings such as (or similar to) "Patent," "Patent Attorney," or "Invention Companies." The invention companies listed have doubtless worked with patents, so they're a good place to start asking for some advice. Try to gain as much knowledge about the patent industry as possible. The remainder of this chapter will give you a good idea of what sort of questions to ask.

Be sure to check professionals for credentials. *Make sure* they're actually registered patent attorneys or registered patent agents. There are many people out there who *call* themselves patent-this or patent-that, or have other fancy titles that allude to the patent process. Don't be fooled: There are only three individuals who can go before the patent office.

1. A patent agent
2. A patent attorney
3. Yourself

Nobody else can file through the United States Patent Office. No one. That's the law.

Also, be sure that you don't show your idea to *anyone* or sign *anything* until you've investigated the area thoroughly. Once you've revealed an unprotected idea to a company or marketing firm, and you fail to protect it for one year, it becomes public property, and all your rights to it are lost. There are a lot of scams out there that will try to suck you in.

The Cost of Inventiveness

Once you've found a patent attorney or agent with whom you feel comfortable, the next step is to hire him or her. Realize right here and now that, for the first time, your brilliant idea is going to start costing you real money. Usually, you can get started on the patent process for between two and four thousand dollars. One the other hand, obtaining patents on certain inventions may cost many thousands more; it all depends on the complexity of your brainchild. Think long and hard before you make that investment—but if you're absolutely, positively sure you've got a product that's really worth something, proving it to everyone is worth the time and effort. Not to mention the money.

Disclosure Documents

If several thousand dollars is too rich for your blood, there is another way in which you can still protect yourself. As a service provided to inventors, the Patent Office accepts and preserves for a two-year period papers disclosing an invention. This disclosure, while not in any way reflecting patent ownership, is accepted as evidence of the dates of conception of the invention and the identity of the inventor. Disclosure documents are destroyed after two years, unless referred to in a separate letter in a related patent application.

A minimal fee of $6 must accompany the disclosure, which is limited to written matter or drawings on paper or other thin, flexible material (such as linen or plastic drafting material). It must have dimensions of, or be folded to dimensions not to exceed, 8½" x 13". Each page should be numbered. Text and drawings should be of such quality as to permit reproduction. Photographs are acceptable.

The disclosure must also be accompanied by a stamped, self-addressed envelope and a duplicate copy. Both the original and the duplicate copy must be signed by the inventor. The papers will be stamped with an identifying number, and returned with the reminder that the disclosure document may be relied upon *only* as evidence of the date of conception, and that a patent application *must* be filed in order to provide patent protection.

CHAPTER 3

The Patent Process

At last you've got both your lawyer and your prototype, and you're ready to start filing. But don't hold your breath: A patent can take up to two years to obtain. And during that period, your application may even be rejected several times on the basis of its claims—or simply on what you are claiming is unique about your invention. We'll discuss claims in detail later in this chapter, but for now you should know that if the Patent Office rejects any of your claims, you have the option of reworking them, through your agent or attorney, and resubmitting them. (Of course, each time you resubmit there may be additional filing and agent's or lawyer's fees. Again, you've *really* got to believe in the product to go through all of this.

What Is a Patent?

Now that you're committed, let's take a closer look at exactly what a patent is, where it came from, and what it can and can't do for you and your invention.

In the simplest of terms, a *patent* is a grant of an intellectual property right by the government to an individual. The individual can be the inventor, his or her heirs, or a third party (also called an "assign") to whom the inventor has delegated his or her rights. The grant of a patent is made exclusively through the United States Patent and Trademark Office. Once the patent on an invention has been granted (provided certain maintenance fees have been paid) protection of your invention under it lasts for seventeen years.

The rights conferred by a U.S. patent apply only throughout the United States and its territories and possessions. What are those rights? In the words of the grant itself: "...*the right to exclude others from making, using or selling the invention throughout the United States.*"

Notice that what is granted is *not* the right to make, use, or sell—you *already* have that right. What is granted *is* the right to stop someone else from doing the same with your idea. Any person is ordinarily free to make, use, or sell anything he pleases. A grant from the government to do so is (fortunately) not necessary. Since the patent does not grant the right to make, use, or sell the invention, the patentee's own right to do so is dependent upon the rights of others and on whichever general laws might apply.

Rights Not Granted

It is important to note that just because someone has received a patent for an invention, he or she does not have the right to make, use, or sell that invention if doing so violates any laws. For example, even the inventor of a new, patented cigarette lighter would not be entitled to use his or her own cigarette lighters without first making certain that they conformed to state and federal product safety codes. A patentee may *not* make, use, or sell his or her invention if:

- Its sale is forbidden by a law.
- Doing so would infringe on the prior rights of others.
- Doing so would violate federal antitrust laws; such as by resale price agreements or entering into combination in restraints of trade.
- Doing so would violate the pure food and drug laws.

Ordinarily, though, there is nothing that prohibits a patentee from making, using, or selling his or her own invention, unless he or she thereby infringes upon another patent still in force.

Right of Exclusion

Since the essence of the right granted by a patent is that of excluding others from commercial exploitation of the invention, the patentee is the *only* person who may make, use, or sell the invention. Others may not do so without authorization from the patentee. The patentee may manufacture and sell the invention himself or he or she may license the right, that is, give authorization, to others to do so.

United States Patent Laws: A Brief History

In 1641, the first patent on the North American continent was granted to Samuel Winslow by the General Court of Massachusetts, for his new method of making salt.

The Founding Fathers, when they designed the Constitution in 1787, wanted to ensure that the newborn central government would be committed to continually fostering domestic technology. To that effect, Article I, Section 8, decreed that:

> Congress shall have the power...to promote the progress of science and useful arts, by securing for limited times to authors and inventors the exclusive rights to their respective writings and discoveries.

Under this article, Congress enacted various laws relating to the granting of patents. The first U.S. patent law was enacted in 1790, and the first U.S. patent went to Samuel Hopkins of Vermont, for his new method of making carbonate of potassium, a chemical used in glassmaking. In the 1800s they were considering shutting down the patent office, on the assumption that everything that *could* be invented already *had* been invented. In 1930, plants became patentable! In 1931, Henry Bosenberg obtained the first plant patent for a climbing rose. A general revision of the law was enacted on July 19, 1952. It came into effect on January 1, 1953, and is codified in Title 53 of the United States Code. This is the law now in effect.

Patent law specifies exactly what can be patented, and under what conditions. It also establishes the Patent and Trademark Office as the administrator of all laws relating to patents.

The U.S. Patent and Trademark Office: Yesterday

As a distinct bureau, the Office is generally considered to date from 1802, when a separate official with the State Department, who later became known as the Superintendent for Patents, was placed in charge of all patents. An 1836 revision of the patent laws reorganized the Office and designated the official in charge to be the Commissioner of Patents and Trademarks.

During its history, the office has undergone several major moves. It remained in the State Department until 1849, when it was transferred to the Department of the Interior. Then in 1925 it was transferred again, this time to the Department of Commerce—where it has remained until this day. In 1967, the Patent Office physically moved to its current headquarters in the Crystal Plaza, a complex on U.S. Highway 1, between Washington, D.C., and Alexandria, Virginia.

The U.S. Patent and Trademark Office: Today

In addition to administering all intellectual property laws, the Patent and Trademark Office performs a variety of related duties, including:

- The examination of patent applications
- The granting of patents
- The publication of a listing of all issued patents, as well as various informative publications concerning patents and the patent process
- The maintenance of a public patent search room so anyone may examine the records of existing patents
- The supplying of copies of patent records and other papers, on request.

Obviously the Office is a great data base for would-be inventors, historians, and students. Issuing roughly 70,000 patents a year, the Patent Office records contain a wealth of valuable, curious, and fascinating information.

The Office does *not* have jurisdiction over questions of infringement or the enforcement of patents, nor over matters relating to the promotion or utilization of patents or inventions. This is important to understand. The purpose of the Patent Office is to *grant* and *record* patents; beyond those functions it has no legal authority. It would be both inappropriate and pointless to approach the Patent Office, should someone infringe on your patent. It is up to the court system to uphold your patent, should someone try to steal your idea.

How the Patent Office Works

The Commissioner

The head of the Office is referred to by the rather lengthy title of Assistant Secretary and Commissioner of Patents and Trademarks. His staff includes a Deputy Assistant Secretary, a Deputy Commissioner, several assistant commissioners, and a host of lesser officials. As head of the Office, the Commissioner technically performs all the official duties regarding the granting of patents and the registration of trademarks. He or she also prescribes the rules for the conduct of the proceedings in the Office, and the recognition of patent attorneys and agents. These rules, however, are subject to the approval of the Secretary of Commerce.

In addition to these duties, the Commissioner is also responsible for making decisions on specific patent issues and questions when these matters are brought before him or her by petition. He or she is, in short, very busy.

Examiners

The work of examining the patent applications is divided among a number of examining groups, each headed by a group director and staffed by a number of examiners. Each of these groups also has jurisdiction over certain assigned fields of technology.

It is the individual examiner's job to review your patent application and determine whether or not your invention will be granted a patent. Examiners are also responsible for identifying *conflicting applications*—in other words, more than one application from different inventors for the same invention. In such cases, the examiner initiates a special proceeding, called an *interference*, to determine which applicant was the original inventor.

Board of Patent Appeals

When a patent is *not* granted, an appeal to reconsider the application can be taken to the Board of Patent Appeals. Even if the board rejects the application, it is possible to get an interference on their decision by submitting a petition. If the interference is granted, the matter will be reviewed and decided by the Commissioner. The Commissioner's decisions are final.

Additional Services

Other units of the Patent Office also perform various services—such as receiving and distributing mail, receiving new applications, handling sales of printed copies of patents, making copies of records, inspecting drawings, and recording assignments.

At present, the Patent and Trademark Office has about 3,100 employees, of whom about half have technical and legal training. Patent applications are received at the rate of over 100,000 per year. The Office receives over five million pieces of mail each year.

What *Can* Be Patented?

In the words of the statute: "[Any person who] invents or discovers any new and useful process, machine, manufacture, or composition of matter, or any new and useful improvements thereof, may obtain a patent."

This is a short quote, but an important one in terms of understanding the patent laws, so let's take a closer look at the language used, and consider what some of its terms mean.

Process in this context refers to a mode, or specific method, of performing a task. It includes everything from arranging a production assembly line to a chemical process that can turn gold into lead.

Machine is used here in the general sense of a man-made tool.

Manufacture refers to any and all products or articles that are fabricated, fashioned, or otherwise built.

Composition of matter relates specifically to chemical compositions. The category includes mixtures of ingredients (such as a cookie recipe) as well as new chemical compounds.

Useful in this context means that the idea seeking a patent must (a) serve some purpose and (b) operate in conjunction with that purpose. In other words, an invention claiming to be an eggbeater but that only fried steaks could *not* be patented as an eggbeater. A machine that did not do what it was supposed to do would not be considered "useful" in this context.

What *Can't* Be Patented?

Throughout the history of its law, the patent statute has undergone various interpretations in the courts. In the process, certain limits on what is *not* patentable have been determined. Several examples follow.

Business Methods and Printed Matter

The courts have held that methods of doing business and printed matter cannot be patented.

Atomic Weaponry

The Atomic Energy Act of 1954 excludes, for obvious reasons, the patenting of inventions useful only in the utilization of special nuclear material or sources of energy for atomic weapons.

Certain Mixtures

In the case of ingredient mixtures, such as medicines, a patent cannot be granted unless more is produced by the mixture than just the basic effects of its components. In other words, if you mixed aspirin and a decongestant, and the results merely relieved headaches and runny noses, the medicine would *not* be patentable. If, however, the mixture cured hives, it would be doing something new, and therefore *could* be patented. So-called patent medicines are, in fact, normally *not* patented. Theirs is the same sort of ruse marketers use when calling a product "all natural" when it simply isn't.

Ideas

A patent also cannot be obtained on a "mere" idea or suggestion. The patent is granted on the new machine, process, or mixture itself—*not* on the idea for it. A complete description, including blueprints, and possibly also a working prototype, will be required in order for the submission to be patented.

Previous Public Use (One-Year Limit)

Patent law further provides that an invention cannot be patented if, during the full year prior to the date of the patent application:

- It has already been known to, or used by, others in this country.
- It has already been patented.
- It has been described in a printed publication in this or a foreign country.

Let's look at this carefully. If, for more than one year before you file a patent application for your invention, the same invention was on sale, in public use, *or even described in a publication issued anywhere in the world*, your patent application will be rejected.

It doesn't matter when the invention was made—the patent application will still be rejected. In fact, even if it is yourself, the inventor, who first describes or uses the invention in public, you must apply for the patent *within one year*, or your right to it will be lost and the invention will automatically fall into the public domain, where *anyone* can freely use it.

Similarities to Existing Inventions

A patent may be refused if the slight differences between the "new" invention and an existing one are too obvious or are "shown by the prior art." *Prior art* is another way of saying that the innovation is implicit in existing technology. Your invention must be different enough from anything that has already been used or described that it would not appear to be an obvious variation by someone having only ordinary skill in the related area of technology.

Suppose your invention is for a type of alarm clock that

plays a cassette tape instead of a radio. Since both parts of the invention already exist, and most anyone could come up with the idea of combining the two, your patent application would be rejected on prior art grounds.

The patent may sometimes also be rejected even if the new invention is *not* directly "shown by the prior art." For instance, neither an umbrella made of plastic (the substitution of one material for another) nor a "giant-size" coat hanger (a change in size) are patentable.

Who May Apply for a Patent?

According to the law, only the inventor may apply for a patent. There are, of course, certain exceptions, but by-and-large, if a person who is *not* the inventor applies for the patent, it is, if obtained, invalidated, *and* the applicant is subject to criminal penalties. The only time that someone other than the inventor is allowed to apply for the patent is when one of the following conditions is met:

1. The inventor is dead.
2. The inventor is insane.
3. The inventor cannot be found.
4. The inventor refuses to apply for the patent.

If the inventor is dead, the application may be made by his or her legal representative, which is to say, the administrator or executor of the estate. If the inventor is insane, the patent application may be made by a legal guardian. If an inventor refuses to apply for a patent, or cannot be located, a joint inventor, or a person having a proprietary interest in the invention, may apply on behalf of the stubborn or missing creator.

Joint Patents

If two or more persons participate in the creation of the invention, they may apply for a patent together, as *joint* inventors. A person who makes only a financial contribution cannot be considered a joint inventor. If a joint inventor is erroneously omitted, or someone is wrongly named as a joint inventor, the application can still be corrected.

Though it may seem unfair, officers and employees of the Patent and Trademark Office are prohibited by law from applying for a patent, and from acquiring—directly *or* indirectly, *except* by inheritance or bequest—any patent, or even any right or interest in any patent. So if you want to be an inventor, don't get a job at the Patent Office!

Conflicting Inventors

In the case of more than one applicant for the same invention, the patent will be granted to whoever is deemed the "prior" inventor. The terms "conception of the invention" and "reduction to practice" often are encountered in connection with such questions of priority.

Conception of the invention refers to the devising of the means for accomplishing the invention.

Reduction to practice refers to the actual construction of the invention in physical form. In the case of a machine it includes the building of the device, and in the case of an article or composition it includes the making of it—and so on. The filing of a patent application that completely discloses the workings of the invention is considered the equivalent of a reduction to practice. The inventor who proves to be *both* the first to conceive of the invention *and* the first to reduce it to practice will be held to be the prior inventor. More complicated situations, which however go beyond the scope of this book, can of course arise.

CHAPTER 4

The Patent Application

As you'll soon realize, the patent application itself is a highly complex, strictly structured document that requires the assistance of a qualified patent attorney or patent agent to complete. It can be broken down into the following parts:

- A written specification consisting of
  A description
  Claims
  An oath or declaration
- A drawing, in those cases in which a drawing is necessary
- The filing fee

The patent application is not forwarded to an examiner until all its required parts are received. If the papers and parts are incomplete, or so defective that they cannot be accepted, you will be notified about the deficiencies and given a time period in which to correct them. In such instances, a surcharge may be required. If you don't respond within the time period, the application will be either returned or disposed of. When an application is refused as incomplete, the filing fee may be refunded—but a handling fee will be charged.

You should make sure that all the parts of your application are deposited to the Office together. If you deliver it piecemeal, each separate part must be signed and accompanied by a letter that clearly connects it to the remaining parts. It is much more efficient, and certainly less confusing, to submit an application all at once.

When the completed application is received, it is numbered in serial order. You are given a *filing receipt*, which informs you of the serial number and filing date. The *filing date* is the date on which a specification (including claims) and any required drawings are received in the Office—or, in the case of a previously incomplete or defective application, is the date on which the last part completing or correcting the application is received.

Neatness Counts

The "specification" and "oath or declaration" may be legibly written or printed in *permanent ink* on one side of a sheet of paper. The Office *prefers* typewriting on legal- or letter-size paper, 8 by 11½ to 10½ by 13 inches, typed or written 1½-spaced or double-spaced, with a left and top margin of one inch. If the papers filed are not correctly, legibly, and clearly written, the Patent and Trademark Office may require type-written or printed papers.

No Money-Back Guarantees

Once the completed application is submitted, neither the papers pertaining to it nor the filing fee will be returned for any purpose whatsoever. If the applicant does not have copies of the papers, the Office will furnish copies—for a fee.

The Application Itself

Now, let's take a detailed look at each part of the application.

The Oath or Declaration

Either an oath or a declaration by the applicant is required by law. You, as the inventor, must swear that you believe yourself to be the original and first inventor of the "subject matter" of the application—in other words, of the invention. An *oath* must be sworn to before a notary public, or any other such officer authorized to administer oaths.

A *declaration*, on the other hand, need *not* be notarized.

Under the following circumstances, a simple "declaration" may be used instead of an oath:

- As part of the original application for a patent involving designs, plants, and other patentable inventions
- When a patent is reissued
- When the applicant is making new claims on an invention already applied for
- When the applicant is filing a divisional or continuing application

The oath or declaration must be signed either by the inventor personally or by the person entitled by law to make the application on the inventor's behalf. As a further homage to the precision of the process, the full first or middle name of each inventor, without abbreviation, and a middle or first initial, is also required.

Blank application forms are *not* supplied by the Patent and Trademark Office, but most patent attorneys or agents will be able to provide you with some blank forms—perhaps for a small fee. You can also refer to the blank patent application forms that appear in the appendixes of this book.

The Specification

The specification consists of the following, in this order:

1. Title of the invention—or, a preamble stating the name, citizenship, and residence of the applicant *plus* the title of the invention may be used
2. Cross-reference to related applications, if any
3. Brief summary of the invention
4. Brief description of all views of the drawings, if there *are* drawings
5. Detailed descriptions
6. Claim or claims
7. Abstract of the disclosure

Title

If the title of the invention, which should be as short and specific as possible, does not appear at the beginning of the application, it should appear as a heading on the first page of the specification.

Clear Language

It is absolutely required that the specification be written in such full, clear, concise, and exact terms that anyone skilled in the technical area to which the invention pertains, or with which it is most nearly connected, will be able to make a working version of the invention.

Distinguishing the Invention

The specifications must set forth the invention in a way that distinguishes it from other inventions and from what is "old," or in common use. As we have said before, it cannot simply describe an idea or principle; it must describe a specific embodiment of that idea or principle. If applicable, it must also explain the mode of operation. The manner in which the inventor intends to carry out his or her invention (the methodology, construction, etc.) must also be set forth.

Variation on Existing Inventions

If your invention is actually an improvement on an older one, the specifications must point out the particular parts of the older invention that are being improved. The description should refer *only* to the specific improvement, and to those parts of the old invention which (a) work directly in conjunction with the new invention, or (b) may be necessary to a complete understanding of the new one.

Technical Abstract

On a separate page, immediately after the claims and in a separate paragraph, a brief abstract of the technical disclosure must be set forth under the heading, "Abstract of the Disclosure."

Summary

Prior to the detailed description there must also be a brief summary of the invention, indicating its nature and substance—and possibly including a statement of its objective. This summary must be completely consistent with any other descriptions of the invention that appear throughout the application.

Description of Drawings

If applicable, a brief description of the various views contained in the drawings must also be included. Each distinct view of the invention, and of parts of the invention in the view, should be numbered separately. In the detailed description, you must then refer to the views and the parts by specifying the appropriate reference numbers.

Claims

The specifications must conclude with one or more claims. These claims particularly point out the exact subject matter which you, the applicant, regard as the invention. *The claims are the operative part of the patent.* Novelty and patentability are judged by the courts on the basis of the claims.

Each part of your invention that performs a separate function warrants a separate claim. With something as simple as a new type of paper clip, for example, there may be as few as two or three claims. Without actually citing actual claims from the patent, you would not want to say that the paper clip is merely a bent piece of metal configured in such a way as to hold pieces of paper together. You would want to say how the metal is bent, how much tension is required, types of materials other than metal that may be used as well as a range of sizes, metal thickness and types and strengths of metal.

In terms of your protection, the claims are probably the most important part of the patent application. There's an old adage in the patent business that says if your application goes through too quickly, you've probably made your claim too weak. "Weak" in this sense means *too narrow*—in which case you set your idea up for either easy theft or the circumvention of your patent.

Should it come time for a judge to determine whether or not someone has infringed on your patent, you want the exact nature of your invention to be as clear to the court as possible. So try, with the help of your lawyer, to make your claims as tough, broad, and specific as you possibly can. The stronger your claims are, the less likely it is that someone will be able to come along and steal or circumvent the idea.

For purposes of the application, the claims should be very brief, but precise. Any unnecessary details should be eliminated—but you must also be careful to include all the essential features necessary for the examiner to distinguish your invention from what is old.

Multiple Dependent Claims

When you present more than one claim, you can organize them in what's known as a *multiple dependent* manner. In other words, one claim may refer back to one or more others, further honing and restricting the information presented in the earlier claims.

If your application uses multiple dependent claims, latter claims must include a reference naming the former claim (or claims) they refer to. The reference must also specify the further limitation that the latter claim proposes. A multiple dependent claim is understood by the Patent Office to incorporate all the limitations of that particular claim. Thus one multiple dependent claim cannot be the basis for another.

Conformity

The claim or claims must completely conform to the descriptions of the invention as set forth in the remainder of the specification. The terms and phrases used in the claims must have enough basis in the terminology of the description to ensure that their meanings will be understood whenever reference is made to the description.

Drawings (Including Standards for Drawings)

Patent applicants are required by law to furnish a drawing of their invention with their application whenever the nature of the invention makes this possible. This stipulation encompasses practically all inventions except compositions of matter and processes. Even so, some sort of drawing may well be useful even in the case of processes.

The patent drawing is *not* a simple sketch. It must show *every* feature of the invention specified in the claims. Naturally, the drawings must also be such that they can easily be understood. The Office, as you may have guessed by now, also requires that the drawing be in a particular form.

The size and type of the paper used, the margins, the ink, and various other details all are carefully specified by the

Office. While there may be enough detail in this process to make you cross-eyed, the reason for specifying drawing standards is quite simple. It is so that all submitted drawings can be printed and published in a uniform style when the patent is issued.

Lest the reader think we exaggerate the detail and complexity of Office rules, we now present, unedited, the specifications for drawings as they appear in Title 37 of the Code of Federal Regulations:

1.84 Standards for drawings.

(a) Paper and ink. Drawings must be made upon paper which is flexible, strong, white, smooth, non-shiny and durable. India ink, or its equivalent in quality, is preferred for pen drawings to secure perfectly black solid lines. The use of white pigment to cover lines is not normally acceptable.

(b) Size of sheet and margins. The size of the sheets on which drawings are made may either be exactly 8½ by 14 inches (21.6 by 35.6 cm.) or exactly 21.0 by 29.7 cm. (DIN size A4). All drawing sheets in a particular application must be the same size. One of the shorter sides of the sheet is regarded as its top.

(1) On 8½ by 14 inch drawing sheets, the drawings must include a top margin of 2 inches (5.1 cm.) and bottom and side margins of 1/4 inch (6.4 mm.) from the edges, thereby leaving a "sight" precisely 8 by 11¾ inches (20.3 by 29.8 cm.). Margin border lines are not permitted. All work must be included within the "sight". The sheets may be provided with two 1/4 inch (6.4 mm.) diameter holes having their centerlines spaced 11/16 inch (17.5 mm.) below the top edge and 2¾ inches (7.0 cm.) apart, said holes being equally spaced from the respective side edges.

(2) On 21.0 by 29.7 cm. drawing sheets, the drawing must include a top margin of at least 2.5 cm., a left side margin of 2.5 cm., a right side margin of 1.5 cm., and a bottom margin of 1.0 cm. Margin border lines are not permitted. All work must be contained within a sight size not to exceed 17 by 26.2 cm.

(c) Character of lines. All drawings must be made with drafting instruments or by a process which will give them satisfactory reproduction characteristics. Every line and letter must be durable, black, sufficiently dense and dark, uniformly thick and well defined; the weight of all lines and letters must be heavy enough to permit

adequate reproduction. This direction applies to all lines however fine, to shading, and to lines representing cut surfaces in sectional views. All lines must be clean, sharp, and solid. Fine or crowded lines should be avoided. Solid black should not be used for sectional or surface shading. Freehand work should be avoided wherever it is possible to do so.

(d) Hatching and shading. (1) Hatching should be made by oblique parallel lines spaced sufficiently apart to enable the lines to be distinguished without difficulty. (2) Heavy lines on the shade side of objects should preferably be used except where they tend to thicken the work and obscure reference characters. The light should come from the upper left-hand corner at an angle of 45°. Surface delineations should preferably be shown by proper shading, which should be open.

(e) Scale. The scale to which a drawing is made ought to be large enough to show the mechanism without crowding when the drawing is reduced in size to two-thirds in reproduction, and views of portions of the mechanism on a larger scale should be used when necessary to show details clearly; two or more sheets should be used if one does not give sufficient room to accomplish this end, but the number of sheets should not be more than is necessary.

(f) Reference characters. The different views should be consecutively numbered figures. Reference numerals (and letters, but numerals are preferred) must be plain, legible and carefully formed, and not be encircled. They should, if possible, measure at least one-eighth of an inch (3.2 mm.) in height so that they may bear reduction to one twenty-fourth of an inch (1.1 mm.); and they may be slightly larger when there is sufficient room. They should not be so placed in the close and complex parts of the drawing as to interfere with a thorough comprehension of the same, and therefore should rarely cross or mingle with the lines. When necessarily grouped around a certain part, they should be placed at a little distance, at the closest point where there is available space, and connected by lines with the parts to which they refer. They should not be placed upon hatched or shaded surfaces but when necessary, a blank space may be left in the hatching or shading where the character occurs so that it shall appear perfectly distinct and separate from the work. The same part of an invention appearing in more than one view of the drawing must always be designated by the same character, and the same character must never by used to designate different parts. Reference signs not

mentioned in the description shall not appear in the drawing, and vice versa.

(g) Symbols, legends. Graphical drawing symbols and other labeled representations may be used for conventional elements when appropriate, subject to approval by the Office. The elements for which such symbols and labeled representation are used must be adequately identified in the specification. While descriptive matter on drawings is not permitted, suitable legends may be used, or may be required in proper cases, as in diagrammatic views and flow sheets or to show materials or where labeled representations are employed to illustrate conventional elements. Arrows may be required, in proper cases, to show direction of movement. The lettering should be as large as, or larger than, the reference characters.

(h) [Reserved]

(i) Views. The drawing must contain as many figures as may be necessary to show the invention; the figures should be consecutively numbered if possible in the order in which they appear. The figures may be plain, elevation, section, or perspective views, and detail views of portions of elements, on a larger scale if necessary, may also be used. Exploded views, with the separated parts of the same figure embraced by a bracket, to show the relationship or order of assembly of various parts are permissible. When necessary, a view of a large machine or device in its entirety, may be broken and extended over several sheets if there is no loss in facility of understanding the view. Where figures on two or more sheets form in effect a single complete figure, the figures on the several sheets should be so arranged that the complete figure can be understood by laying the drawing sheets adjacent to one another. The arrangement should be such that no part of any of the figures appearing on the various sheets are concealed and that the complete figure can be understood even though spaces will occur in the complete figure because of the margins on the drawing sheets. The plane upon which a sectional view is taken should be indicated on the general view by a broken line, the ends of which should be designated by numerals corresponding to the figure number of the sectional view and have arrows applied to indicate the direction in which the view is taken. A moved position may be shown by a broken line superimposed upon a suitable figure if this can be done without crowding, otherwise a separate figure must be used for this purpose. Modified forms of construction can only be shown in separate figures. Views should not

be connected by projection lines nor should center lines be used.

(j) Arrangement of views. All views on the same sheet should stand in the same direction and, if possible, stand so that they can be read with the sheet held in an upright position. If views longer than the width of the sheet are necessary for the clearest illustration of the invention, the sheet may be turned on its side so that the top of the sheet with the appropriate top margin is on the right-hand side. One figure must not be placed upon another or within the outline of another.

(k) Figure for Official Gazette. The drawing should, as far as possible, be so planned that one of the views will be suitable for publication in the Official Gazette as the illustration of the invention.

(l) Extraneous matter. Identifying indicia (such as the attorney's docket number, inventor's name, number of sheets, etc.) not to exceed 2³/4 inches (7.0 cm.) in width may be placed in a centered location between the side edges within three-fourths inch (19.1 mm.) of the top edge. Authorized security markings may be placed on the drawings provided they are outside the illustrations and are removed when the material is declassified. Other extraneous matter will not be permitted upon the face of a drawing.

(m) Transmission of drawings. Drawings transmitted to the Office should be sent flat, protected by a sheet of heavy binder's board, or may be rolled for transmission in a suitable mailing tube; but must never be folded. If received creased or mutilated, new drawings will be required. (See 1.152 for design drawing, 1.165 for plant drawings, and 1.174 for reissue drawings.)

The requirements relating to drawings are strictly enforced, but a drawing not complying with all of the regulations may be accepted for purpose of examination, and correction or a new drawing will be required later.

Applicants are advised to employ competent draftsmen to make their drawings.

Sheet Identificaton

No names or other identifications are permitted "within" the drawing. Applicants are expected to use a small space above and between the hole-punch locations to identify each sheet. This identification may consist of the attorney's name and docket number, or the inventor's name and case number. It may also include the sheet number and the total number of sheets.

Complying With the Rules

Quite a mouthful, eh? However, though these requirements are strictly enforced, a drawing *not* complying with all the regulations *will* be accepted for purpose of examination; correction or a new drawing will be required *later*.

Obviously, it is advisable to employ a competent draftsman to make your drawings.

Models, Exhibits, and Specimens

Since the description of the invention in the specification and the drawings alike must be sufficiently full and complete to allow someone with the proper understanding thereof to construct a working version of the invention, models are *not* required in most patent applications. In fact, models generally are not admitted unless specifically requested by the examiner—and the examiners do not request them often. On the other hand, if the invention is an alleged perpetual-motion device, a working model may well be requested.

Compositions of Matter

When the invention is a composition of matter, the applicant may be required to furnish specimens of the composition, or of its ingredients or intermediates, for inspection or experiment. If the invention is a microbiological invention, a deposit of the microorganism involved is required.

Allowance and Issue of Patent

Once every "i" has been dotted and every "t" crossed, your patent application will ultimately appear before an examiner from the appropriate group. If, either on the initial examination of the application or during the reconsideration of the application, a patent is found to be allowable, a notice of allowance will be sent to either the applicant or the applicant's attorney or agent. A sample of a patent appears as Patent Appendix V.

Fees

A fee for issuing the patent is due within three months after the date of the allowance notice. The fee for each original or reissue utility patent is $1,130. The issue fee for a design application is $400. The amount of the fee is reduced by one-half for small entities. If timely payment is not made, the application will be regarded as abandoned. However, a provision in the statute provides that if the applicant can prove an unavoidable delay, the Commissioner may accept the fee late.

Once the fee is paid, the patent is issued as soon as possible after the date of payment. (The actual time involved will vary, depending on the volume of printing that the Patent Office has on hand.) The patent grant then is either delivered or mailed directly to the inventor. On the date of the grant, the patent file becomes open to the public, and printed copies of the specification and drawing are made available to anyone requesting such.

National Defense Exception

In case the publication of an invention by the granting of a patent would be detrimental to the national defense, patent law gives the Commissioner the power to withhold the grant, and to order that the invention be kept secret for such period of time as the national interest requires.

Maintenance Fees

Any utility patent issued from an application filed on or after December 12, 1980, is subject to the payment of maintenance fees. These fees must be paid in order to keep the patent in force. Maintenance fees are due at $3\frac{1}{2}$, $7\frac{1}{2}$, and $11\frac{1}{2}$ years from the date the patent is granted. The fees can be paid without a surcharge during a "window period," the six months preceding each due date (e.g., three years to three years and six months, etc.). An exact fee schedule is listed in Appendix IX.

Failure to pay the current maintenance fee on time may result in expiration of the patent. A six-month grace period immediately following the due date is provided, during which the maintenance fee may be paid, but with a surcharge. The Patent and Trademark Office does *not* mail to patent owners notices that maintenance fees are due. If, however, the maintenance fee isn't paid on time, efforts *are* made to remind the patentee that the fee and surcharge may still be paid during the grace period.

Patents relating to certain pharmaceutical inventions may be extended by the Commissioner for up to five years, to compensate for marketing delays due to federal premarketing regulatory procedures. Patents relating to all other types of inventions can be extended only by the enactment of special federal legislation.

Expiration

After the patent has expired, the rights to the invention fall into the public domain. Then, *anyone* may make, use, or sell the invention without the permission of the patentee, provided that matter covered by any *unexpired* patents is not used. The terms of a patent may not be extended except by a special act of Congress!

Correction of Granted Patents

Once the patent is granted, it is no longer within the jurisdiction of the Patent and Trademark Office—with the following exceptions:

Clerical Errors: The Office may issue, without charge, a certificate correcting a clerical error it has made in the patent, when the printed patent does not correspond to the record in the Office. (These are mostly corrections of typographical errors made in printing.)

Applicant Errors: Some minor errors of a typographical nature made by the applicant may be corrected by filing a certificate of correction. A charge is assessed.

Applicant Disclaimer: The patentee may disclaim one or more claims of the patent by filing a disclaimer.

Reissue Patents: When the patent is defective in certain respects, the law provides that the patentee may apply for a reissue patent. This is a patent granted to replace the original—but only for the balance of the unexpired term. The nature of the changes that can be made by means of the reissue are rather limited, and new matter cannot be added.

Patent Examination: Any one may file a request for examination of a patent, along with the required fee, on the basis of prior art consisting of patents or printed publications. At the conclusion of the proceedings, a certificate setting forth the results of the reexamination proceeding is issued.

Patent Marking

A patentee who makes or sells patented articles, or a person who does so either for or under the patentee, is required to mark the articles with the word "Patent" *and the number of* the patent. The penalty for failure to mark is that the patentee may not recover damages from an infringer unless the infringer was duly notified of the infringement and continued to infringe after the notice.

The marking of an article as patented when it is not in fact patented is against the law, and subjects the offender to a penalty.

Other Types of Patents

The type of patent we've been discussing up until now is also referred to as a *utility* patent. Most inventions fall under this category, but there are two other types of patents you should be aware of—*design patents* and *plant patents*. Many patent companies and attorneys will use only the word "Patent," so you should make certain exactly which type of patent is being discussed.

Design Patents

Patent laws provide for the granting of design patents for any new, original, and ornamental design for an article of manufacture. A design patent protects only the *appearance* of an article, *not* its structure or utilitarian features. The filing fee for each design application is $280; the issue fee is $400. A design patent has a term of fourteen years, and *no* fees are necessary to maintain a design patent in force.

If, on examination, it is determined that an applicant is entitled to a design patent under the law, a notice of allowance will be sent to the applicant (or the applicant's attorney or agent) calling for the payment of an issue fee.

The drawing of the design patent conforms to the same rules as those governing other drawings, but no reference characters are required. The specification of a design application is short, and usually follows a set form. Only *one* claim is permitted.

Design Patent Cautions

Many invention companies and some unethical lawyers push design patents so they can make a fast buck. These patents are relatively inexpensive, and comparatively easy to get. However, the design patent is *not* as effective as a utility patent in terms of overall protection. It protects only the *design* of an item, and designs can be easily changed. So think long and hard before you go after a design patent!

If you have a particular shape in mind for something (a horseshaped automobile, for example) which would be a long-term product involving a considerable amount of money, then you might *want* to consider obtaining a design patent. But if you want to invent, let's say, a different calculator, it's a little risky simply to obtain a design patent—because then anyone can change the design and simply use basically *your* calculator. The design patent doesn't really afford you *any* protection in that instance. Obviously the patent to go after is the utility one.

Plant Patents

The law also provides for the granting of a patent to anyone who has invented or discovered and "asexually" reproduced a distinct and new variety of plant. This category includes cultivated mutants, hybrids, and newly found seedlings. This does *not* include a tuber-propagated plant, or a plant found in an uncultivated state.

Asexually propagated plants are those reproduced by means other than seeds—such as by the rooting of cuttings, layering, budding, grafting, and so on. As for the restriction on tuber-propagated plants, the term "tuber" is here used in a very narrow horticultural sense. It means a short, thickened portion of an underground branch. The *only* plants covered by the term "tuber-propagated" are the Irish potato and the Jerusalem artichoke.

An application for a plant patent consists of the same parts as other applications, but various special rules apply. The application papers for a plant patent and any responsive papers must be filed in duplicate, but only one need be signed (generally the original); the second copy may be a legible copy of the original. The duplicate file is sent to the Agricultural Research Service, Department of Agriculture, for an advisory report on the plant variety.

The specification should include a detailed description of the plant and the characteristics that distinguish it from both known related varieties and its antecedents. The description must be expressed in the general form followed in standard botanical textbooks or other publications dealing with the varieties of the kind of plant involved (evergreen tree, dahlia plant, rose plant, apple tree, etc.), and not as a broad non-botanical characterization such as is commonly found in nursery or seed catalogs.

The specification should also include the origin or parentage of the new plant, and must particularly point out where and in what manner the new variety of plant has been asexually reproduced. If color is a distinctive feature of the plant, that aspect should be positively identified in the specification by reference to a recognized color dictionary. If the new plant variety originated as a newly found seedling, the specification must fully describe the conditions (e.g., cultivation and environment) under which the seedling was found growing, in order to establish that it was *not* found in an *uncultivated* state.

A plant patent is granted on the *entire* plant. It therefore follows that only one claim is necessary—and in fact only one is permitted. The oath or declaration must include the statement that the applicant has *asexually* reproduced the new plant variety.

Although plant patent drawings are *not* mechanical drawings, they should nevertheless be artistically and competently executed. The drawing must disclose all of the distinctive characteristics of the plant capable of visual representation.

When color is a distinguishing characteristic of the new variety, the drawing *must* be in color.

Two duplicate copies of original color drawings must be submitted. Color drawings may be made, in either permanent watercolor or oil—or, in lieu thereof, photographs may be submitted which have been made by color photography. In any case, the paper must correspond in size, weight, and quality to the paper required for other drawings. Mounted photographs are also acceptable.

Specimens of the plant variety, or of its flower or fruit, should *not* be submitted unless specifically called for by the examiner.

The filing fee on each plant application is $460, and the issue fee is $570. For a qualifying small entity, filing and issue fees are reduced by half.

All inquiries relating to plant patents and pending plant patent applications should be directed to the Patent and Trademark Office, *not* to the Department of Agriculture.

It is also possible to protect the ownership of *sexually reproduced* plants. The Plant Variety Protection Act (Public Law 91-577), approved on December 24, 1970, provides for a system of protection under the administration of a Plant Variety Protection Office within the Department of Agriculture. Requests for information regarding the protection of sexually reproduced varieties should be addressed to:

The Commissioner of Plant Variety Protection Office
Agricultural Marketing Service
National Agricultural Library Bldg., Room 500
10301 Baltimore Blvd.
Beltsville, MD 20705-2351

CHAPTER **5**

Licensing and Marketing

The first stage of any new product is its creation. The second stage is protecting it. The third stage is selling it. All three are equally important—and equally difficult. Now that we've covered the length and breadth of the patent process, in this chapter we'll discuss some of the steps you should take to protect and market your invention.

Patent Pending Status

Once your patent is filed, you enter what is known as *Patent Pending* status. (For presentation as a modifying term, we'll normally use the lower-case spelling, as in "patent-pending paper clip.") You're given a registration number from the Patent Office which you can (and should) print on all information you release in conjunction with your invention. In most cases, Patent Pending status allows you to begin marketing and selling your invention with a reasonable amount of safety. Even so, before you begin to market your idea, you must understand that this status does *not* provide complete security. The only *real* protection in this country is to have the patent itself.

Given this situation, it might seem that the better course of action would be to wait until you have the patent before proceeding to the market with your invention. However, since this process, as we've seen, could take several years, we recommend that you *don't* wait. Most legitimate companies that regularly deal with patents and inventors *will* recognize a Patent Pending status.

Invention Company Rip-Offs

Marketing and licensing probably are the most abused stages of the whole invention business: They are loaded with scams and rip-offs. Newspapers and magazines are full of advertisements for so-called invention companies. These outfits will effectively do nothing except take your hard-earned money, and put your idea in severe jeopardy. If the Latin words *caveat emptor* ("let the buyer beware") ever applied, they apply here. Invention companies make their profits *not* by marketing and licensing inventions, but by charging inventors elaborate fees.

Look at it this way: If people earn their living by selling ideas, they'll only be interested in ideas they think they *can* sell. When an idea they believe they *can* peddle shows up, they'll be more than willing to risk their time and effort on it. On the other hand, if people earn their living mostly by *charging fees*, they'll only be interested in *acquiring fees. Any* idea or desperate inventor will do—and we do mean *any*. For a recent television news report, investigators submitted the same idea three separate times over the course of a few months, each time under a different person's name. The "invention" company signed up each one, asking for exorbitant fees to represent it. Its operatives specifically said to each applicant, "No, we never take the same idea from different people. Your idea is unique."

Another thing these companies will do is go out of their way to make you think that your idea, no matter how ridiculous, is absolutely wonderful—the next *wheel*! Many will also want to charge you five to six hundred dollars to do a "research report" on the market for you. The report they will give you is something they keep on computer, pre-prepared, like a form letter. *No* research for your product is actually carried out.

It is estimated that "invention companies" earn between four and eight hundred million dollars a year from these fees. There really isn't much impetus for them to license and market ideas if they're earning that much for doing nothing.

We cannot caution you enough on this. If you have faith in your idea, be patient until you find a professional who *shares* that faith. That's the only way you can feel assured that someone is looking out for your idea.

The Two Options

Now, assuming you've gone through the patent process and you have your Patent Pending status, the next thing you want to know is how to get your product to the market in a safe and effective manner. You'll have two basic options: going it alone (i.e., raising money, and manufacturing and distributing your invention yourself) or attempting to negotiate a deal with another company to do all the work for you. Assuming you'll want to do the latter, we think that your first step should be to acquire an agent. We'll look at going it alone later in this chapter.

Agents

We recommend that you go through a legitimate agent. There are several reasons for this, but two stand out: (1) agents are more expert in the area of selling; and (2) agents have a lot of contacts.

Let's go back to our earlier example of the new paper clip. Now that you have a patent-pending paper clip, you'll want to approach a manufacturing company—but unless yours is a knockout, wonderful paper clip unlike any the world has ever seen, and is guaranteed to earn billions, you'll find that most such companies are very reluctant to deal with you as an individual. Even though you're in a patent-pending status, and even if you do have a unique and terrific invention, it's very tough to get through to a company on your own. As we mentioned earlier, companies are inundated with literally hundreds—even thousands—of ideas every year.

The advantage of working through a license or marketing agent is that, unlike you, he or she will have a track record (a history of sales as well as a professional background in the industry). This gives the agent an important edge. That "pro" will, in general, get more responses and more respect from a company—a *lot* more receptivity.

Attention having been gotten, a lot of the agent's job involves matchmaking. As part of his or her routine, the agent keeps in touch with a number of people who do a variety of things. These people can then match your invention up with the right people in the right places. The agent thus becomes your link in the business loop, a loop that otherwise can prove almost impossible to break into.

Above and beyond these duties, an agent will smooth over all the problems, make sure you're protected, and—in a way—become the parent of your product.

How Do You Find a Good Agent?

This is the same problem we dealt with earlier, regarding patent attorneys. The answer, again, is *shop around*. Agents often are found through word of mouth, references, research, and/or trial and error. Do some interviewing: You are, after all, *hiring* someone. See what different agents have to say. Let them try to *sell themselves* to you. Contact inventors who have worked with them.

As a rule of thumb, if an agent wants a lot of money up front, stay away. We know this advice may get a lot of people in the industry upset, since even some legitimate agencies charge hundreds up front, but we don't believe in doing that. If an agent likes your product and believes in your idea, he or she should be willing to take a chance on it.

Minimal Fees

There may be some minimal fees involved. Much in the same way, a literary agent may charge seventy-five dollars to three hundred dollars to evaluate an author's manuscript. A licensing agent may charge fees to screen useless or silly inventions and to ensure that they will be working with serious clients. Someone who wants to build say a NASA platform out of Jell-O should be discouraged from taking up too much of the agent's time. (*This* fee could be thirty-five to one hundred dollars.) If somebody wants five or ten thousand dollars to market your idea, just say goodbye: That extortioner is not interested in doing anything for *you*.

Another thing you should check on is whether or not the agent has access to legal services—a patent attorney on staff or on retainer is an important indication of how much support an agent will be able to provide for you.

A lot of people suggest checking an agent's record with the Better Business Bureau, but we think that's a waste of time. Companies that are members of the Better Business Bureau have sixty to ninety days to correct complaints. Almost every fraudulent company in the patent business will do everything they can to resolve any complaints lodged with the Better Business Bureau within that time frame, so that no one ever finds out about it. They could be running the biggest fraud going, but as long as their complaint record is clean, people checking through the Bureau will think they're a legitimate business.

In our opinion, the Better Business Bureau is not a great resource. What people don't realize is that it charges businesses a large amount of money to belong to the Bureau. *That's* where the Better Business Bureau makes its money—*not* from consumer advocacy. We are dubious of these types of operations. The reports they send out are so vague that they're ineffectual. The Bureau is certainly something you should include in your research, but you shouldn't rely *entirely* on them.

Nondisclosure Statement

Once you've settled on an agent, make sure you get him or her to sign a nondisclosure or confidentiality agreement before showing that person your invention. This will protect your idea from possible theft.

How Much Do Agents Charge?

An agent generally takes a cut amounting to between 25 and 50 percent of royalties, leaning toward the latter. That may seem a little high, but you have to consider the fact that there are *many* expenses involved in marketing patents. It's very rare that the first company you send something to will want to use your invention, or make you an equitable offer for it. Most products go through ten to twenty different companies before finding the right one.

It's one thing to create an idea, but it's quite another to market it. They are two separate but intertwined processes, yet in some cases it's more difficult to license and sell a product than actually to create it. The bottom line is that the *agent* doesn't make money unless making money for *you*.

Types of Contracts

Once you have your agent, your ultimate goal will be to make a deal with a company for the rights to your inventions. There are two basic types of deals that can be made: You can either sell your rights entirely or you can enter what's known as a *licensing agreement.*

What Is Licensing?

When you *license* a product, you are in essence leasing it to a company for their use—usually in exchange for a royalty payment. In our imaginary example for our new patented paper clip, either you or your agent would approach an interested company, ABC Stationery, and say, "Okay, you can use my idea for this product if you pay me a certain percentage of the price for each and every one that you sell." In the case of licensing, as the inventor you would *still* own all the rights to your paper clip.

Licensing can be beneficial for a number of reasons. First and foremost is the fact that there are no headaches for you. Once you've signed a licensing agreement, you more or less sit back and collect checks. In most cases you're not liable for *anything*—you're not involved in the production, the distribution, or the advertising. And in most cases you can't even be held legally liable for the *product!* Myriad products on the market today are licensed.

Royalties

On a typical licensing agreement, you should expect a 3 percent to 7 percent *royalty,* based on the wholesale price of the item. *Wholesale* is the amount the company will charge retail outlets for the product. So, if a box of Joe's New Paper Clips retails at about $2, the wholesale price would probably be about $1, in which case an inventor's 5 percent royalty would amount to five cents per unit. These percentages can vary—sometimes they're a little lower, sometimes a little higher. But the average range is in the 5 percent to 7 percent area.

Liability

If you have a competent licensing agent, he or she is going to protect you as much as possible in the product agreement set up with a company. Your agent will insert in the agreement what's known as an *indemnification clause.* This will protect you from lawsuits. In short, it will say that you, the inventor, are not *liable* for any problems that arise with relation to your product; the company is. Therefore, you *shouldn't* be apt to be sued.

Advances

If you have a particularly good invention, a hot product which various companies are chomping at the bit to license, very often your agent can negotiate an *advance* for you. An advance is a flat fee paid to the inventor against possible future royalties. It is important to realize that often the advance is not extra money, but part and parcel of the royalty payment. Then again, everything in this business is open to negotiation. It depends on how hot the product is; there are no hard and fast rules.

In our Joe's Paper Clip model, let's suppose Joe is given a thousand-dollar deductible advance from the company. In that case, his product would have to sell twenty thousand units before he would receive any additional royalties. But what, you may ask, if it never sells more—if in fact it sells *less* than that before being taken off the market? Not to worry: Most deductible advances are nonrefundable. If the product fails to earn back the $1,000 advance, you, the patent owner, are *not* liable for the difference.

Often an advance can consist of as much as a year's royalties. But at times a deal can be made with *no* advance, on a *royalty-only* basis—in which case it is important that you trust the company you are doing business with.

For more complete information, refer to the sample licensing contract that appears in Appendix VIII at the end of this Part.

Before Signing Anything

As an inventor it always behooves you, regardless of whether you have an agent, to look into the company you're making a licensing deal with—if only for your own peace of mind. If you've got a deal going with Union Carbide, great! But what if it's with a company that's going bankrupt? Before you sign, investigate the company as thoroughly as possible. If you have access, get a Dunn & Bradstreet report on the company. This will give you a payment history of the company. It will tell you *how* (and *if*) they pay their bills, whether they're in bankruptcy, the significance of any liens against them, and so on.

And of course you can go to one or another local library and look up the company's history and records. Larger libraries are even more wonderful resources with vaster amounts of information. They can also supply you with all sorts of information over the phone.

Many people don't realize how much down-and-dirty business knowledge is available to them, *free of charge*, through common sources. Explore, explore, explore!

You'll also want to be sure that the company is going to advertise, promote, and market your product correctly. You may get a deal, but if the product just sits there, what good is it? You should discuss how and where the product is going to be sold. Will it sell in the U.S., Canada, or in the European market? (Tomorrow the world?)

Very often, in that first contract, you're going to want to limit the marketing area—perhaps just to part of the United States. Don't give away everything in one show! Also, be conscious of your protection. If you only have a United States patent, *don't* sign away foreign rights.

Get everything in writing, and acquire as many guarantees as possible. By *guarantees* we mean stipulations as to how many units they are going to produce, how and when they are going to pay you, and so on.

Selling the Invention Outright

If you sell the patent outright, you sell all the rights to it, and you have nothing left. (This is true in every area of intellectual property—art, music or inventions, whatever.) Although each case is different, we *don't* usually recommend selling the patent outright. In the long run, there is much more money to be made through licensing—unless of course you're offered an incredibly outlandish sum to do otherwise, and actually receive it.

Profiting From Repressed Ideas

Companies often buy an idea just to keep it *off* the market. Let's say you've come up with a new invention for a perpetual-motion machine. Now, this might be a great and wondrous thing for some, but for many it would mean the end of their livelihood—because it probably would put all "nonperpetual" machines of that type perpetually out of business. Machine shops all over the world are aware that this kind of device might, if ever produced, completely destroy their business. So, in response, some of the larger machine companies get together and offer to buy the patent for a hundred million dollars, just to save themselves. With the inventor happily wealthy and out of the picture, they'll proceed to *bury* the idea.

This may sound outlandish, but it happens all the time—particularly in the auto industry. Oftentimes, large auto manufacturers will buy the patents for products that will enhance the workings of an automobile (gasoline-saving products, for example), just to junk them.

This can get quite nasty. Companies threatened by your idea could connive to try to sequester your idea *and you* in the courts. Large companies have potent legal resources; they can challenge your patent and entangle you in litigation for *years*. It's a big, bad world of jealously protective people out there.

Self-Production

Your final option is to produce your invention on your own, which is to say *without* the help of an outside company. In this case, you keep all the *rights* to your patent, but also experience all the *risks*. Of the three options—licensing, selling, or going it alone—this is probably the most risky. Not only will you need funds for the patent process, but you'll also need major start-up capital, which you probably won't be able to raise from your family unless they're *very* well-off.

Self-Manufacture Concerns

Manufacturing isn't simply a matter of taking your product to a factory, building it, and selling it. Many other areas have to be carefully considered—such as:

- Liability
- Hiring employees
- Getting materials
- Federal regulations
- Quality control
- Distribution

Let's touch briefly on these. Suppose you've devised a new type of stapler, and you manufacture and sell it yourself. If someone cuts a finger on your invention, *you*, as the manufacturer, become liable for damages. And just one successful damage suit can cost you millions of dollars.

As another example, let's say you create a new doll, which you want to manufacture and market yourself. Among other things, you have to hire people to do the sewing, get liability insurance, and buy materials that must meet various state and federal regulations. (The cloth for a doll's dress has to be fire-retardant, for example—and the colors of the dye in the fabric have to be safe and nontoxic.)

Right from the start you'll also have to ask yourself a lot of very detailed questions about the safety of your product. If a child puts it in his or her mouth, can he or she get sick? (Is the doll safe in itself?) Are there pieces that can come off? Pieces that can be swallowed? The government has regulations, insurers have regulations, everyone else has regulations, and you'll have to abide by *all* of them.

Another major concern of the self-producer is *distribution*. How are you going to get your product to the marketplace? Are you going to distribute it yourself, or hire a distributor? A distributor is usually going to want to take a percentage of the profits, and probably an up-front fee as well. The distributor may well want a hundred thousand dollars as a safeguard, in case your dolls don't sell. Again, there are a *lot* of hassles in doing it yourself.

Increased Profits

On the positive side, your profits can be much greater if you go it alone. Instead of receiving a 5 percent–7 percent licensing royalty, you could find yourself making 25 percent–50 percent on the value of your product. It is, however, also possible you'll *lose* a ton of money if you go solo. Larger companies, with commensurately larger advertising budgets and distribution networks, can give your product a better chance at being a success, because if the advertising and other promoting of your product isn't done properly, it could easily die on the shelf.

Even if you do the job by yourself and your product is a smashing success, you still have to make sure you can meet the continuing demand for it. Have contingencies set up in case the demand increases rapidly. Too, you have to be prepared to deal with "knock-offs" (duplicates of your product) that will try to take up any slack.

To end where we began, we repeat: Both the rewards and the risks are much greater if you produce your own product.

When It May Be Better to Go It Alone

There are some situations where we recommend you try to produce your own product *despite* all the risks. If you have a very simple invention which can be produced inexpensively, it's actually advantageous to try producing and selling it on your own—but you'll have to start out *really* small.

For example, suppose you own the patent for a simple novelty item that can be put together and packaged for about a dollar and a half or less, per unit. If you can get some family and friends together to assemble a little financing, it would be much better for you to try selling your product on your own than to go with a licensing agreement. If you had a licensing agreement for your simple product, and it was sold for two dollars apiece, you might make as little as three cents a unit. On the other hand, if you made the product yourself, you'd be making fifty cents for each. That's a big difference—and it adds up fast.

Keep in mind that this commentary applies only to the *simplest* of products. If you have a high-tech invention, a new kind of tape recorder, or a revolutionary toothpaste, then things start getting much more complicated. *Then* you'd have to get involved in *all* kinds of regulations and complex manufacturing procedures. In that case, licensing might well be the better choice. Still, it really depends on what you are willing to risk. Talk with other business people, and keep your eyes wide open: There are a lot of people you can learn a lot from.

Starting Small

To give you an idea of how to start small, let's look at food products, even though they generally *can't* be patented. A common way that new food products get on the market is via domestic "chefs" making small quantities of their product and selling all of it themselves. It's a long route to success, albeit for some a fulfilling one. You can start by going to flea markets or bazaars or anywhere else you can get a small place to sell your product cheaply. When someone likes your product, you take the name and address and such, and start building a mailing list. Using your mailing list, you start sending out brochures and doing mail-order business from your home. Once you get the product somewhat established in the local area, you'll start being noticed by bigger companies. Eventually you might even get an offer from one or more to buy you out. This is a very simplified picture—but in general it shows how this kind of business approach works. Famous Amos Cookies, David's Cookies, Aunt Millie's Spaghetti Sauce, Sara Lee, Arnold's Bread, and many other giants started out this way.

One famous self-success story is that of Trivial Pursuit, a game invented by a few Canadian advertising copywriters. Try as they might, they couldn't sell it—*anywhere*. Refusing to give up, they acquired investment capital from friends and ac-quaintances, and started manufacturing it in small numbers themselves. After they had attained *some* success on their own, Selchow and Righter purchased the rights from them, allowing an unheard-of 15 percent royalty. You know the rest of the story.

Another (and even bigger) small-start product was The Teenage Mutant Ninja Turtles, which began as a self-published comic at an initial investment of about two hundred dollars. Some reward, Dude!

We repeat that it is vitally important that you *believe* in your product. You can't just have an idea and expect everything to fall into place that will ensure its success. You must have enough faith in it to keep you willing to work long and hard for its success.

Raising Funds

There are two areas for which you may need to raise funds. The first is to pay both the attorney and the fees necessary for filing your patent application. The second is if you want to produce your product on your own.

In the first instance it's possible that, prior to applying for your patent, you'll be in contact with a company that wants to manufacture your product. The company may offer to pay for your application, in exchange for some of the patent rights. In the long run, of course, it's better to give up as few of your rights as possible. What can you do?

Letter of Intent

One thing you can do is ask the company in question to sign a *letter of intent*. This is a legal document indicating their willingness to market your idea. You can even take a letter of intent to a bank and use it to get a loan to pay for your filing costs. Of course, you'll also have to demonstrate that the product not only is salable but also has good potential.

Borrowing From Friends and Family

The other, and most obvious, way to raise money while maintaining your rights is to approach friends and relatives, and either borrow the money from them outright or offer them small percentages in your product. In the latter case, you shouldn't hand over too much of a percentage, since you're not dealing with a great amount of money—three or four thousand dollars usually is enough to cover the lawyer and filing fees for an average patent.

Patent Attorney as Temporary Partner

Another option is to work out a deal with a patent attorney. Sometimes, in lieu of payment, one will accept a percentage of your profits—but we advise caution in negotiating an agreement of this nature, because sometimes attorneys (like agents) get greedy. *Don't* hand over 10 percent, or even 5 percent, of your profits *in perpetuity* (i.e., forever). Make sure you put a *cap* on the attorney's share. With a cap, the total amount of money an attorney can make from your invention is limited. And you can cut off his or her profits at say twenty-five thousand or one hundred thousand dollars—whatever you agree on. At that point the attorney is paid off and out of the picture.

As you negotiate, understand that this is a terrific deal for the attorney. If the patent's for a good product, it's costing that participant nothing except the filing fee, since the legal expenses are a result of his or her own labor.

Raising Start-Up Funds / Going It Alone

Another situation in which you would need to raise funds is having the patent and wanting to market the product yourself. If you have a patent, you have something salable, something tangible—and if that's *good,* you can raise money for it.

You might want to try for a bank loan, but that's often a bit of a long shot (especially is you have no track record in running a business). There are also a variety of entrepreneurial organizations which can help. Further, check the yellow pages of your phone book, under "Investment Advisory Service" and "Investment Securities." And look in the classified section of your local paper for ads about investment money. You'll be surprised to find that there are scads of people out there who want to invest in various products. On Broadway, they're called "angels."

There actually are organizations that will look at new ideas, and finance them for a percentage of the company they help set up. Others will locate and place you with a company.

Beware of Scams Again

Much as with the so-called invention companies we warned you about earlier, you also have to watch out for the "investment brokers." There are a lot of scams in this area, too. Some may want you to pay them an up-front fee for finding you money. It sounds bizarre, but they'll actually try to charge you five hundred or a thousand dollars to find money for you— something you can basically do yourself. A *legitimate* broker or investment counselor will usually take only a *small* percentage of the money you receive. For example, if you receive a hundred thousand dollars, the "legit" broker may take 5 percent as a fee. Working on a mutually agreeable percentage basis gives brokers a strong incentive to do a good job for you.

Other Fund-Raising Ideas

You can also be as creative in raising money as you were when you invented the product. Think of what you own that you can sell. Work an extra job. You'll eventually see that how far you're willing to go and how much extra effort you're willing to put in, depends on how much you believe in your product.

Advertising Yourself and Your Product

Before your product is protected, you should discuss it with *as few* people as possible. Once your invention *is* protected, however, the exact *opposite* applies, since then, the more attention you can attract to yourself and your product, the better. At that point, anything at all you can do that will draw attention to yourself is beneficial. The question thus becomes: How do you attract attention?

There are many ways to do that. One of the most popular and most effective is the *press release*. You put together the best small one you can and mail it to, say, a combination of two hundred radio stations, newspapers, and magazines. Even if only one or two will give you coverage—presto! You've got free advertising!

A press release is a brief, but very informative story relating to your invention. For example, you should say who the inventor(s) is, why the product was invented, where it is available, how much it costs, and what makes it so terrific. In short, you want to include as much information as you can, and sell your product, in as little as two to six concise paragraphs. Try to make it as exciting as possible. Hyperbole is allowed and humor helps. Remember, you are not only selling the idea to a reader but also to the potential editor who may broadcast or print this item. Also you should include any photos or endorsements that you may have and, if it is a local venue, then include information about yourself.

For examples of press releases, check your local library for marketing books or simply look in your local newspapers. Much of the business, social and local sections contain numerous rel1eases and are often identified by the absence of a byline (author's name).

Initially send out about two hundred or so to as many magazines, newspapers, newsletters, trade publications, wire services (AP & UPI), and radio and television stations as you can. You may find that smaller outlets are more receptive to the stories, but have no fear, if the release is interesting it will get picked up by the big boys.

But there's much more you can do. Remember, you don't have to go through just the usual channels, calling up company after company and trying to get some secretary to put you through to whoever handles people like you. *Talk it up.* Discuss your product everywhere you go. Give a business card to everyone you meet. Put fliers on cars. Run small classifieds in newspapers. Effective advertising and promotion can mean anything from renting a plane for skywriting, down to pulling a small wagon behind you with a banner attached. If you want to meet the head of a company, find out where he or she eats, lives, plays, and otherwise spends his or her time, and work on getting together at some convenient hour and place. Be at least as creative as you were in conceiving your invention.

Press releases aside, television is probably the *most* effective medium to exploit. There are numerous *local* cable shows that you can try to get yourself on, for starters. As you succeed, keep a scrapbook of your appearances—build a portfolio. Once you have that evidence to show around, it becomes easier to get into some of the larger newspapers, or on some of the more popular TV shows. The fuller your portfolio, the greater your legitimacy in the eyes of *all* media.

In the end, the difference often is *chutzpah.* As long as it's legal and you're not hurting anybody, use it. There are thousands of people out there with absolutely *no* talent, but they're celebrities because they've made *themselves* famous. Much the same is true for products as well.

Selling Your Patent

Now let's backtrack a little and talk in some depth about what it is your patent rights allow you to do that pertains to this area of our considerations. A patent, like a trademark and a copyright, is personal property. It may be sold to others, mortgaged, or bequeathed by a will and passed to the heirs of the deceased patentee. Patents—or even an *application* for a patent—may be transferred or sold, via an *instrument in writing* (also known as a written contract). A contract that *assigns* patent ownership is referred to as *assignment.* Once the assignee becomes the owner of the patent, he or she has the same rights the original patentee had. Properly executed assignments are regarded as absolute until canceled, either by the parties or by the decree of a competent court.

Patent law also allows for the assignment of a *part interest*—that is, a percentage of the patent (half, one-fourth, etc.). Territorial assignments can also be issued that cover only a part of the country. For example, you could sell a half interest in your patent to one person for the states east of the Mississippi and to another for the states to the west.

Patent ownership can also be claimed if the inventor defaults on a mortgage or loan. A mortgage of patent property passes ownership to the mortgagee or lender, until the mortgage has been satisfied. At that point, a retransfer from the mortgagee back to the mortgagor—the borrower—is made.

An assignment, grant, or conveyance of a patent or patent application must be signed before a notary public or some other officer authorized to administer oaths or perform notarial acts. The certificate of such acknowledgment constitutes *prima facie*, or *apparent* evidence of the execution of the assignment, grant, or conveyance.

Recording of Assignments

As one of its duties, the Patent Office keeps a record of assignments and grants. This record serves as legal notice. If an assignment, grant, or conveyance is not recorded in the Office within three months of the date of its assignment, it is *void* against a subsequent purchaser for any money or other form of compensation without notice, unless it is recorded prior to the subsequent purchase.

A patent *assignment* should identify the patent by number and date, the name of the inventor, and the title of the invention. A patent *application assignment* should identify the application by its serial number, date of filing, the name of the inventor, and the title of the invention, all as stated in the application.

Sometimes an application assignment is executed *at the same time* that the application is prepared—before it has even been filed in the Office. In that case, the assignment must adequately identify the application by its date of execution, the name of the inventor, and the title of the invention, so that there can be no mistake as to the application intended.

If an application has been assigned and the assignment is recorded *on or before* the date on which the issue fee is paid, the patent will be issued to the *assignee* as owner. If the assignment is a part interest only, the patent will be issued to the inventor *and* the assignee as *joint* owners.

Joint Ownership

Any joint owner of a patent, no matter how small the part interest, may make, use, and sell the invention for his or her own profit, *without* regard to the other owner. He or she may also sell his or her interest in the patent, or any part of it, or grant licenses to others, without regard to the other joint owner, *unless* the joint owners have made a contract governing their relation to each other. It is *dangerous* to assign a part interest in your patent without a definite agreement! The agreement should specify concerning *all* the parties' respective rights, *and* their obligations to each other.

Granting License

In addition to assigning rights, the owner of a patent may also grant licenses to others. Since the patentee has the right to exclude others from making, using, or selling the invention, no one else may do any of these things without his or her permission. A *license*, then, is the permission granted by the patent owner to another to make, use, or sell the invention. No particular form of license is required; a license may include whatever provisions the parties agree upon, including the payment of royalties, etc.

The drawing-up of a license agreement (as well as assignments) is within the purview of the ordinary general attorney at law, but even that level of counselor should also be familiar with "local" patent matters: A few states have prescribed certain formalities to be observed in connection with the sale of patent rights.

Patent Infringement

The best likelihood is that once you have an established patent and are marketing your invention, no one will try to steal your invention. If it is a particularly successful invention, though, you can be assured that others will try to imitate it without infringing on your patent. Sometimes, however, they will cross the line, and you, to protect your rights, may well have to take them to court.

Legally speaking, *patent infringement* consists of the un-authorized making, using, or selling of a patented invention within the territory of the United States during the term of the patent. If your patent *is* infringed upon, you may sue for relief in the appropriate federal court. You may ask the court for both of the following:

- An injunction to prevent the continuation of the infringement
- An award of damages because of the infringement

In such an infringement suit, there are two common defense strategies you should be aware of. The defendant will most likely raise the question of the validity of your patent, claiming perhaps that it is not really a new idea, or is one which falls in the public domain. If you are suing, the court will decide on the validity of your patent.

The defendant may also claim that what he or she is doing does *not* constitute infringement. Here is where your carefully worded claims become important. Infringement is determined primarily by the language of those claims and, if what the defendant is manufacturing or using *does not* fall within the language of any of the claim of the patent, there is *no* infringement.

Suits for infringement of patents follow the standard rules of procedure for the federal courts. From the decision of the district court, there is an appeal to the Court of Appeals for the Federal Circuit. The Supreme Court may thereafter take a case by "writ of certiorari," which essentially asks permission to appear before the high court. If the United States Government should infringe on your patent, you'll be able to sue for damages in the United States Claims Court. The government is legally allowed to use *any* patented invention without permis-sion of the patentee, but the patentee is in turn entitled to obtain compensation for the use of his or her idea by or for the government.

If the patentee notifies anyone who is infringing upon the patent or if the patentee threatens suit, the one charged with infringement may start the suit in a federal court.

As we stated earlier, the Patent Office has no jurisdiction over questions relating to patent infringement. In examining patent applications, no determination is made as to whether the invention sought to be patented infringes on any prior patent. An improvement to an invention may be patentable, but it might infringe on a prior unexpired patent for the invention improved upon, if there is one.

Conclusion

You are now armed with enough information to enable you to successfully navigate your idea through the patent process and out into the marketplace—where, with some luck, it will earn money for you for years to come.

We strongly suggest that you review the additional information that appears in the Patent Appendixes. The publication lists, sample forms, and sample contracts can provide invaluable guidance.

As a closing thought for this section, we'd like to point out that each and every reader is a potential inventor. Many of us never go anywhere near the whole route, though, and some of us even run out of steam right after talking our idea over with some family and friends. Oddly enough, particularly in the United States, many people *do* go the whole route. For example, in 1993, out of about 188,099 applications, 107,332 patents were granted. This means that about one in every 3,000 people in the United States owned their own patents then. In fact, in many ways the United States is itself a unique invention—so it should be no surprise that so many ideas originate here. Good luck!

An Inventive Tale

It was the 1940s and the United States was embroiled in World War II. Many individuals and companies stateside were asked to give their assistance to the war effort. Scrap-iron drives were common—but another substance in great demand was rubber, which was needed in everything from tires for jeeps to the soles of soldiers' shoes. The U.S. War Production Board approached General Electric to research and produce an inexpensive rubber substitute. General Electric gave the project to a company engineer named James Wright.

Using silicone oil and boric acid, Wright managed to invent (in 1944) a rubberlike compound with some wondrous properties. It stretched better, bounced higher, and withstood a wider range of temperature than did standard rubber. Unfortunately, it had no industrial advantage over synthetic rubber, and so remained little more than a curiosity. However, in 1945, GE sent samples out to some of the world's leading engineers, asking if they could come up with a use for it. Nothing much came of this effort, but then GE was approached by a shopkeeper named Paul Hodgson, who bought a large mass of the strange stuff from them and started selling it in his store. He put a bit of the blob inside a plastic egg—and named it Silly Putty. And that's the way the ball bounced.

Patent Appendix I

Publications of the Patent and Trademark Office

The Patent Office contains a wealth of information for the would-be inventor. What follows is a list of publications available from them.

Patents

The specifications and accompanying drawings of all patents are published and made available to the general public on the day the patent is granted. To date, over 5,200,000 patents have been issued. A printed copy of any patent, identified by its patent number, may be purchased from the Patent and Trademark Office at a cost of $1.50 each, including postage. Color copies of plant patents may be had at $10.00 each.

It is also possible to order patents that haven't been issued yet, by specifying a subject matter of interest and prepaying a deposit and service charge. This would give you, in effect, a subscription to the patents. For the cost of such a subscription service, a separate inquiry should be sent to the Patennnt and Trademark Office.

The *Official Gazette* of the United States Patent and Trademark Office

The *Official Gazette* of the United States Patent and Trademark Office is a journal relating to patents and trademarks. It has been published weekly since January 1872 (replacing something called the "Patent Office Reports"). It is currently issued in two parts every Tuesday. One part describes patents, the other trademarks. The *Gazette* also contains:

- A claim and selected drawings from each patent granted on that day

- Notices of patent and trademark suits
- Indexes of patents and patentees
- A list of patents available for license or sale
- General information such as orders, notices, changes in rules, changes in classification, etc.

Since July 1952, the illustrations and claims of the patents have been arranged in the *Official Gazette* according to their subject matter, permitting ready reference to patents in any particular field. The street addresses of patentees have been published since May 24, 1960, and a geographical index of residences of inventors has been included since May 18, 1965.

The *Official Gazette* is sold either by subscription or single copies. It can be ordered from:

The Superintendent of Documents
U.S. Government Printing Office
Washington, DC 20402

Copies of the *Official Gazette* may also be found in public libraries of larger cities.

Index of Patents: This annual index of the *Official Gazette* is currently in two volumes: an index of patentees and an index, by subject matter, of the patents. It is also sold by the Superintendent of Documents.

Index of Trademarks: This is an annual index of registrants of trademarks. It is sold by the Superintendent of Documents.

Manual of Classification: This is a loose-leaf book which lists all the classes and subclasses of inventions in the Patent and Trademark Office classification systems. It includes a subject matter index, and other information relating to classification. Substitute pages are issued from time to time. An annual subscription includes the basic manual and substitute pages. It is sold by the Superintendent of Documents.

Classification Definitions: This contains any changes in the classification of patents, as well as definitions of new and revised classes and subclasses. There are currently over 400 classes and over 120,000 subclasses. It is sold directly by the Patent and Trademark Office.

Title 37 Code of Federal Regulations: This reproduction of the patent law includes rules of practice for patents, trademarks, and copyrights. It is available from the Superintendent of Documents.

Basic Facts About Trademarks: This is a general-purpose booklet that contains information for the layman about applications for, and registration of, trademarks and service marks. Copies may be purchased from the Superintendent of Documents.

Directory of Registered Patent Attorneys and Agents Arranged by States and Countries: This is an alphabetical and geographical listing of patent attorneys and agents registered to practice before the U.S. Patent and Trademark Office. It is sold by the Superintendent of Documents.

Manual of Patent Examining Procedure: This is a loose-leaf manual which serves primarily as a detailed reference work on patent examining practice and procedure for the Patent and Trademark Office's Examining Corps. A subscription service includes the basic manual, quarterly revisions, and change notices. It is sold by the Superintendent of Documents.

The Story of the United States Patent Office: This is a chronological account of the development of the U.S. Patent and Trademark Office and our patent system. It also chronicles inventions that have had unusual impact on the American economy and society. It is sold by the Superintendent of Documents.

Patent Appendix II

Correspondence With the Trademark Office

Any and all business with the Patent and Trademark Office can be transacted by writing to:

Commissioner of Patents and Trademarks
Washington, DC 20231

Applicants and attorneys are required to conduct their business with decorum and courtesy. Papers presented in violation of this requirement will be returned. You should also be sure to include your full return addresses, including zip code. The principal location of the office is Crystal Plaza 3, 2021 Jefferson Davis Highway, Arlington, Virginia. The personal attendance of applicants at the Office is unnecessary.

One Letter per Subject

Separate letters (not necessarily in separate envelopes) should be written for each distinct subject of inquiry, such as assignments, payments, orders for copies of patents or records, requests for other services, etc. None of these should be included with letters responding to Office actions in applications.

Correspondence Regarding Applications

When a letter concerns a patent application, it must include the serial number, filing date, and Group Art Unit number. When a letter concerns a patent, it must include the name of the patentee, the title of the invention, the patent number, and the date of issue.

Application Privacy

Patent applications are not open to the public. No information concerning them can be released without the written authority of the applicant, his or her assignee, or his or her attorney, or when it is necessary to the conduct of the business of the Office.

Public Information

The following information is, however, open to the public:

- Patents and related records
- The records of any decisions involving the patents
- The records of assignments other than those relating to assignments of patent applications
- Books and any other records and papers in the Office

These may be inspected in the Patent and Trademark Office Search Room, or copies may be ordered.

The Patent Office cannot:

- Respond to inquiries concerning the novelty and patentability of an invention in advance of the filing of an application
- Give advice as to possible infringement of a patent
- Advise as to the propriety of filing an application
- Respond to inquiries as to whether or to whom any alleged invention has been patented
- Act as an expounder of the patent law or as counselor for individuals, except in deciding questions arising before it in regularly filed cases

Information of a general nature may be furnished either directly or by calling attention to an appropriate publication.

Ordering Copies of Assignments

An order for a copy of an assignment must give the book and page or reel and frame of the record, as well as the name of the inventor; otherwise, an additional charge is made for the time consumed in making the search for the assignment.

Patent Appendix III

The Search Room and Depository Libraries

The Scientific Library of the Patent and Trademark Office at Crystal Plaza 3, 2021 Jefferson Davis Highway, Arlington, Virginia, has over 120,000 volumes of scientific and technical books in various languages, about 90,000 bound volumes of periodicals devoted to science and technology, the official journals of 77 foreign patent organizations, and over 12 million foreign patents.

Patent Searches

Since a patent is not always granted when an application is filed, many inventors attempt to make their own investigations before applying, by examining the patent information available in either the Search Room of the Patent and Trademark Office, or any of the Patent Depository Libraries located throughout the United States. Patent attorneys or agents may be employed to make a so-called preliminary search through prior patents in order to discover if a particular invention or one similar to it has been shown in some prior patent. This search is not always as complete as that made by the Patent and Trademark Office during the examination of an application. The Patent and Trademark Office examiner may, and often does, reject claims in an application on the basis of prior patents or publications not found in the preliminary search.

The Search Room

A Search Room is provided where the public may search and examine any and all United States patents granted since 1836. Patents are arranged according to the Patent and Trademark Office classification system. The Search Room contains a set of

United States patents arranged in numerical order, and a complete set of the *Official Gazette*. The Search Room is open from 8 A.M. to 8 P.M., Monday through Friday, except on federal holidays.

Files Information Room

Here, the public may inspect the records and files of issued patents and other open records. Applicants and their attorneys or agents may inspect their own cases here. Public records may be inspected in the Scientific Library, Search Room, or Files Information Room. Applicants, their attorneys or agents, and the general public are not entitled to use the records and files in the examiners' rooms.

Patent Search by Mail

Those who cannot come to the Search Room may order from the Patent and Trademark Office copies of lists of original patents or of cross-referenced patents contained in the sub-classes comprising the field of search, or they may inspect and obtain copies of the patents at a Patent Depository Library.

Patent Depository Libraries

The Patent Depository Libraries (PDLs) receive all current issues of U.S. Patents and maintain collections of earlier issued patents. The scope of these collections varies from library to library, ranging from patents of only recent years to all or most of the patents issued since 1790.

These patent collections are open to public use. In addition, each PDL offers all the publications of the U.S. Patent Classification System, and other patent documents and forms, and provides technical staff assistance to aid the public in gaining effective access to the information contained in the patents. The collections are organized in patent number sequence.

Computer Data Base

Also available in all PDLs is the Classification and Search

Support Information System (CASSIS), the computer data base. With various modes within a program, it allows the user to search appropriate classifications, and provides the numbers of the patents assigned in order to allow the user to access a copy of the patents in a numerical file of patents. CASSIS also provides the current classification(s) of all patents, permits word searching on classifications, abstracts, and the *Index*, and provides certain bibliographic information on recently issued patents.

PDL List

Due to variations in the scope of patent collections among PDLs, and their hours of service to the public, anyone contemplating the use of a particular library is advised to contact that library, in advance, about its collection and hours, so as to avoid possible inconvenience.

State	*Name of Library*
Alabama	Auburn University Libraries
	Birmingham Public Libraries
Alaska	Anchorage Municipal Libraries
Arizona	Tempe: Noble Library, Arizona State Univ.
Arkansas	Little Rock: Arkansas State Library
California	Los Angeles Public Library
	Sacramento: California State Library
	San Diego Public Library
	Sunnyvale: Patent Information Clearinghouse
Colorado	Denver Public Library
Connecticut	New Haven: Science Park Library 60
Delaware	Newark: University of Delaware Library
Dist. of Columbia	Washington: Howard University Libraries
Florida	Ft. Lauderdale: Broward County Main Library
	Miami-Dade Public Library
	Orlando: Univ. of Central Florida Libraries
Georgia	Atlanta: Price Gilbert Memorial Library, Georgia Institute of Technology
Idaho	Moscow: University of Idaho Library

Illinois	Chicago Public Library
	Springfield: Illinois State Library
Indiana	Indianapolis–Marion County Public Library
Iowa	Des Moines: State Library of Iowa
Kentucky	Louisville Free Public Library
Louisiana	Baton Rouge: Troy H. Middleton Library, Louisiana State University
Maryland	College Park: Engineering and Physical Sciences Library, Univ. of Maryland
Massachusetts	Amherst: Physical Sciences Library, Univ. of Massachusetts
	Boston Public Library
Michigan	Ann Arbor: Engineering Transportation Library, Univ. of Michigan
Minnesota	Minneapolis Public Library and Info. Center
Missouri	Kansas City: Linda Hall Library
	St. Louis Public Library
Montana	Butte: Montana College of Mineral Science and Technology Library
Nebraska	Lincoln: Engineering Library, Univ. of Nebraska–Lincoln
Nevada	Reno: Univ. of Nevada–Reno Library
New Hampshire	Durham: Univ. of New Hampshire Library
New Jersey	Newark: Public Library
	Piscataway: Library of Science and Medicine at Rutgers University
New Mexico	Albuquerque: Univ. of New Mexico Library
New York	Albany: New York State Library
	Buffalo and Erie County Public Library
	New York Public Library (The Research Libraries)
North Carolina	Raleigh: D. H. Hill Library, North Carolina State University
Ohio	Cincinnati and Hamilton County, Public Library of
	Cleveland Public Library
	Columbus: Ohio State University Libraries
	Toledo/Lucas County Public Library

Oklahoma	Stillwater: Oklahoma State University Library
Oregon	Salem: Oregon State Library
Pennsylvania	Philadelphia, Free Library of Pittsburgh, Carnegie Library of University Park: Pattee Library, Pennsylvania State University
Rhode Island	Providence Public Library
South Carolina	Charleston: Medical University of South Carolina Library
Tennessee	Memphis and Shelby County Public Library and Information Center Nashville: Valderbilt University Library
Texas	Austin: McKinney Engineering Library, University of Texas at Austin College Station: Sterling C. Evans Library, Texas A & M Univ. Dallas Public Library Houston: The Fondren Library, Rice Univ.
Utah	Salt Lake City: Marriott Library, University of Utah
Virginia	Richmond: Virginia Commonwealth Univ. Library
Washington	Seattle: Engineering Library, Univ. of Washington
Wisconsin	Madison: Kurt F. Wendt Library, Univ. of Wisconsin–Madison Milwaukee Public Library

Facilities for making paper copies from either microfilm in reader-printers or from the bound volumes in paper-to-paper copies are generally provided for a fee.

Patent Appendix IV

Patent Fee Schedule

The exact amount of the fees are subject to change without notice. Current fees are effective as of May, 1992.

e fee schedule as set forth below contains the fees for utility, design and plant applications to ich may be added certain additional charges for claims, depending on their number. An issue is required when a patent is to be granted. These amounts are reduced by 50% when the plicant is a small entity -- that is an independent inventor, a nonprofit organization, or a small siness concern. An applicant must submit a verified statement to establish status as a small ity. In addition, maintenance fees are due 3 1/2, 7 1/2, and 11 1/2 years after the patent issued keep it in force. **Fees are subject to change annually each October.** Fees may be confirmed ore submitting to the U.S. Patent and Trademark Office.

PATENT FEES

Filing Fees	Fee	Small Entity Fee if applicable
Basic Filing Fee - Utility	690.00	345.00
Independent claims in excess of three	72.00	36.00
Claims in excess of twenty	20.00	10.00
Mutiple dependent claim	220.00	110.00
Surcharge-Late filing fee or oath or declaration	130.00	65.00
Design filing fee	280.00	140.00
Plant filing fee	460.00	230.00
Reissue filing fee	690.00	345.00
Reissue independent claims over original patent	72.00	36.00
Reissue claims in excess of 20 and over original patent	20.00	10.00
Non-English specification	130.00	

Issue Fee

Utility issue fee	1,130.00	565.00
Design issue fee	400.00	200.00
Plant issue fee	570.00	285.00

Maintenance Fees
Applications filed on after December 12, 1980

Due at 3.5 years	900.00	450.00
Due at 7.5 years	1,810.00	905.00
Due at 11.5 years	2,730.00	1,365.00
Surcharge-Late payment within 6 months	130.00	65.00
Surcharge after expiration	600.00	

Patent Service Fees

Printed copy of patent w/o color, regular service	3.00
Printed copy of patent w/o color, expedited local service	6.00
Printed copy of patent w/o color, ordered via EOS, expedited	25.00
Printed copy of plant patent, in color	12.00
Copy of utility patent or SIR, with color drawings	24.00
Certified or uncertified copy of patent application as filed, regular service	12.00
Certified or uncertified copy of patent application, expedited local service	24.00
Certified or uncertified copy of patent-related file wrapper and contents	150.00
Certified or uncertified copy of document, unless otherwise provided	25.00
For assignment records, abstract of title and certification, per patent	20.00
Disclosure Document Program	10.00

For a complete list of other fees contact
the Commissioner of Patents and Trademarks, Washington, D.C. 20231

OTHER ADDITIONAL FEES

Reissue filing fee	690.00	345.00
Reissue independent claims over original patent	72.00	36.00
Reissue claims in excess of 20 and over original patent ..	20.00	10.00
Non-English specification..	130.00	

Extension Fees

Extension for response within first month	110.00	55.00
Extension for response within second month	350.00	175.00
Extension for response within third month	810.00	405.00
Extension for response within fourth month	1,280.00	640.00

Appeals/Interference Fees

Notice of appeal	260.00	130.00
Filing a brief in support of an appeal	260.00	130.00
Request for oral hearing	220.00	110.00

Issue Fees

Utility issue fee	1,130.00	565.00
Design issue fee	400.00	200.00
Plant issue fee	570.00	285.00

Miscellaneous Fees

Extension of term patent	1,000.00	
Requesting publication of SIR - Prior to examiner's action	790.00*	
Requesting publication of SIR - After examiner's action	1,580.00*	
Certificate of correction	70.00	
For filing a request for reexamination	2,180.00	
Statutory Disclaimer	110.00	55.00

Patent Petition Fees

Petitions to the Commissioner, unless otherwise specified	130.00	
Petition to institute a public use proceeding	1,310.00	
Petition to revive unavoidable abandoned application	110.00/	55.00
Petition to revive unintentionally abandoned application	1,130.00/	565.00

Maintenance Fees:
Applications filed on or after December 12, 1980

Due at 3.5 years...	900.00/	450.00
Due at 7.5 years...	1,810.00/	905.00
Due at 11.5 years...	2,730.00/	1,365.00
Surcharge - Late payment within 6 months	130.00/	65.00
Surcharge after expiration	600.00	

PCT Fees - National Stage

Surcharge - Late filing fee or oath or declaration	130.00/	65.00
English translation - after twenty months	130.00	

IPEA - U.S.	620.00/	310.00
ISA - U.S.	690.00/	345.00
PTO not ISA or IPEA	920.00/	460.00
Claims meet PCT Article 33(1)-(4)-IPEA - U.S.	90.00/	45.00
Claims - extra independent (over three)	72.00/	36.00
Claims - extra total (over twenty)	20.00/	10.00
Claims - multiple dependent	220.00/	110.00
For filing with EPO or JPO search report	800.00/	400.00

PCT Fees - International Stage

Transmittal fee	190.00
PCT search fee - no U.S. application	600.00†
Supplemental search per additional invention	160.00†
PCT search - prior U.S. application	400.00†
Preliminary examination fee - ISA was the U.S.	440.00†
Preliminary examination fee - ISA not the U.S.	650.00†
Additional invention - ISA was the U.S.	140.00†
Additional invention - ISA not the U.S.	220.00†

PCT Fees to WIPO

Basic fee (first thirty pages)	490.00*
Basic supplemental fee (for each page over thirty)	10.00*
Handling fee	150.00*
Designation fee per country	119.00*

PCT Fees to EPO

International search	1,320.00

† Effective December 27, 1991.

* WIPO fees subject to periodic change due to fluctuations in exchange rate. Refer to Patent Official Gazette for current amounts.

Patent Service Fees	**Fee**
Printed copy of patent w/o color, regular service	3.00
Printed copy of patent w/o color expedited local service	6.00
Printed copy of patent w/o color, ordered via EOS, expedited service	25.00
Printed copy of plant patent, in color	12.00
Copy of utility patent or SIR, with color drawings	24.00
Certified or uncertified copy of patent application as filed, regular service	12.00
Certified or uncertified copy of patent application, expedited local service	24.00
Certified or uncertified copy of patent-related file wrapper and contents	150.00
Certified or uncertified copy of document, unless otherwise provided	25.00
For assignment records, abstract of title and certification, per patent	20.00

Library Service	50.00
List of U.S. patents and SIRs in subclass	3.00
Uncertified statement re satus of maintenance fee payments	10.00
Copy of non-U.S. document	12.00
Comparing and Certifying Copies, Per Document, Per Copy	25.00
Additional filing receipt, duplicate or corrected due to applicant error	20.00
Filing a Disclosure Document	10.00
Local delivery box rental, per annum	50.00
International type search report	35.00
Self-service copy charge, per page	0.25
Recording each patent assignment, agreement or other paper, ·per property	40.00
Publication in Official Gazette	20.00
Labor charges for services, per hour or fraction thereof	30.00
Unspecified other services	AT COST
Retaining abandoned application	130.00
Handling fee for incomplete or improper application	130.00
Automated Patent System (APS-text) terminal session time, per hr.	40.00
Handling fee for withdrawal of SIR	130.00
Patent coupons	3.00

Patent Enrollment Fees

Admission to examination	290.00
Registration to practice	100.00
Reinstatement to practice	15.00
Copy of certificate of good standing	10.00
Certificate of good standing - suitable for framing	20.00
Review of decision of Director, Office of Enrollment and Discipline	120.00
Regrading of Examination	120.00

The following publications are sold, and the prices for them fixed, by the Superintendent of Documents, Government Printing Office, Washington, D.C. 20402, to whom all communications respecting the same should be addressed:

Official Gazette of the United States Patent Office:

Annual subscription, domestic (non-priority)	$516.00
Annual subscription, (priority)	$687.00
Annual subscription, foreign international Postal Zone	$645.00
Single numbers	$30.00
Annual Index Relating to Patents, price varies	
Manual of Classifications of Patents	$76.00
Foreign	$95.00
Attorneys and Agents Registered to Practice Before the U.S. Patent Office	$21.00
37.Code of Federal Regulations, Part 0-17	$15.00
Manual of Patent Examining Procedure	$78.00
Foreign	$97.50

THE ABOVE PRICES ARE SUBJECT TO CHANGE WITHOUT NOTICE.

All payment of money required for Patent and Trademark Office fees should be made in United States specie, Treasury notes, national bank notes, post office money orders or postal notes payable to the Commissioner of Patents and Trademarks, or by certified checks. If sent in any other form, the Office may delay or cancel the credit until collection is made.

Patent Appendix V

Foreign Patent Protection

Treaties and Foreign Patents

Since the rights granted by a United States patent extend only throughout the territory of the United States, they have no effect in a foreign country. Any inventor wishing to protect his or her inventions in other countries must apply for a patent in each country. Almost every country has its own patent laws, and inventors must make their applications in accordance with the requirements of that country.

In most foreign countries:

- Publication of the invention before the date of application will bar the right to a patent.
- Maintenance fees are required.
- It is required that the patented invention must be manufactured in that country within a certain period— usually three years. If there is no manufacture within this period, the patent may be voided. In most countries the patent may be subject to the grant of compulsory licenses to any person who may apply for a license.

The Paris Convention

There is also a patent treaty, adhered to by ninety-three countries (including the United States), known as the Paris Convention for the Protection of Industrial Property. It provides that each treaty member guarantee to the citizens of the other treaty members the same rights in patent and trademark matters that it gives its own citizens.

In the case of patents, trademarks, and industrial design, the treaty also provides for the right of priority. This means that on the basis of an application filed in one member country the applicant may, within a certain period of time, apply for protection in *all* the other member countries. These later (after-the-fact) applications will be regarded as if they had been filed on the same day as the first application. The period of time within which the subsequent applications may be filed in the other countries is twelve months in the case of applications for patent, and six months in the case of industrial designs and trademarks.

Under these terms, it is possible that later applicants will have legal priority over earlier applications for the same invention! For example, say that applicant A patents electric shoes in the United States in March 1995, and applicant B patents the same product in France in February 1996. If applicant A is the first inventor, and applies for a patent on the shoes in France within twelve months (in March 1996), his or her application will have priority status over B's.

Moreover, these later applications will not be invalidated by acts accomplished in the interval—such as, for example, publication or exploitation of the invention, the sale of copies of the design, or the use of the trademark.

The Patent Cooperation Treaty

Another treaty, known as the Patent Cooperation Treaty, was negotiated at a diplomatic conference in Washington, D.C., in June of 1970 and came into force on January 24, 1978. This treaty facilitates the filing of applications for patent on the same invention in member countries. It provides, among other things, centralized filing procedures and a standardized application format, creating in essence an *international* patent.

The timely filing of an international application gives inventors an international filing date in each country designated in the international application. It also provides:

(1) a search of the invention
(2) a later time period within which the national applications for patent must be filed.

The treaty is presently adhered to by 41 countries, including the United States.

International Patent Attorneys

A number of patent attorneys specialize in obtaining patents in foreign countries. In general, an inventor should be satisfied by the thought of perhaps at best making a marginal profit from foreign patents, or that there is some other particular reason for obtaining them, before attempting to apply for them. They are very expensive.

Foreign Patents Sometimes Require a U.S. License

If your invention was created in the United States and you haven't yet filed for a U.S. patent, or it is less than six months since you did file, then in order for you to submit a foreign patent application you must obtain a license to do so. This would come from the Commissioner of Patents and Trademarks.

The filing of a patent application constitutes an *automatic* request for the foreign license. The granting or denial of such a request is indicated in the filing receipt mailed to each applicant. Six months past the United States filing, a license is no longer required—unless the invention has been ordered kept secret. If it *has* been so ordered, the consent to the filing abroad must be obtained from the Commissioner of Patents and Trademarks during the period in which the secrecy order is in effect.

Foreign Applicants for United States Patents

The patent laws of the United States allow no discrimination on the basis of citizenship; noncitizens also may apply for a patent. There are, however, some special details of the law that pertain to foreign applicants:

- In many other countries, the signature of the inventor and an oath of inventorship are not necessary for the patent application. In the United States, however, the application must be made by the inventor, who (with certain exceptions) must also sign the oath or declaration.
- A United States patent can't be obtained if the invention has already been patented in another country (by the inventor or his or her legal representatives or assigns) via

an application filed more than twelve months prior to the U.S. filing. (In the case of a *design* patent, the time allotted is *six* months.)

• If a patent application is filed in the United States by someone who has previously filed for the same patent in a country adhering to the Paris Convention (see above), the U.S. application will be treated as if it had been filed on the same date as the foreign application—provided that the application in the U.S. is filed within twelve months (again, six months in the case of a design patent) from the earliest date on which the foreign application was filed. A copy of the foreign application, certified by the country's patent office, is required to secure this right of priority. If any application for patent has been filed in any foreign country, the applicant must, in the oath or declaration accompanying the application, state the country in which the earliest such application has been filed, giving the date of filing the application. Any applications filed more than a year before the U.S. filing must also be listed in the oath or declaration.

Oaths in Foreign Countries

When the applicant is in a foreign country, his or her oath or affirmation may be made before a diplomatic or consular officer of the United States, or any other officer having an official seal authorizing the administering of such oaths in that country. The legitimacy of such authority can be proven via the certificate of a diplomatic or consular officer of the United States. The oath must be attested to in all cases by the proper official mark of a seal of the officer before whom the oath is made.

When an *oath* is taken before a U.S. officer in a foreign country, all of the application papers (except the drawing) must be attached together. A ribbon must be passed one or more times through all the sheets of the attached application, and the ends of the ribbon brought together under the seal before the latter is affixed and impressed. Alternately, each sheet must be impressed with the official raised seal of the officer before whom the oath was taken. (When a *declaration* is used, the ribboning procedure is not necessary—nor is it even necessary to appear before an official in connection with the making of a declaration.)

If the application is filed by the legal representative of a deceased inventor, that stand-in must make the oath or declaration. A foreign applicant may be represented by *any* patent attorney or agent registered to practice before the United States Patent and Trademark Office.

A Sample Patent

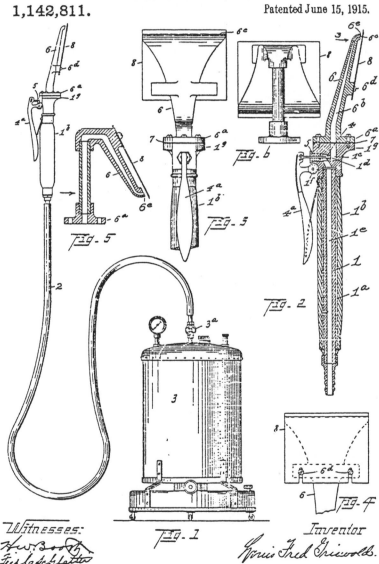

L. F. GRISWOLD.

DEVICE FOR STRIPPING WALL PAPER.

APPLICATION FILED MAY 28, 1914.

1,142,811.

Patented June 15, 1915.

UNITED STATES PATENT OFFICE.

LOUIS FRED GRISWOLD, OF CLEVELAND, OHIO.

DEVICE FOR STRIPPING WALL-PAPER.

1,142,811. Specification of Letters Patent. **Patented June 15, 1915.**

Application filed May 28, 1914. Serial No. 841,554.

To all whom it may concern:

Be it known that I, LOUIS FRED GRISWOLD, a citizen of the United States, residing at Cleveland, in the county of Cuyahoga and State of Ohio, have invented a certain new and useful Device for Stripping Wall-Paper, of which the following is a specification.

My invention relates to the means of removing wall paper from the walls. It is well known to those familiar with the art, that in order to obtain the best results in the redecorating of rooms, it is desirable to remove all the old paper from the wall before applying the new. In some cities there is an ordinance which provides that the wall paper in all rooms which have been occupied by sick people, particularly by people with contagious diseases, shall be removed and the rooms re-papered. In many instances there are several layers of paper on the wall, and the difficulty of removing one or more thicknesses of paper from the wall is well known to those familiar with the art, and the means commonly used for the purpose.

Machines heretofore used for this purpose, and using steam for a loosening medium, have applied the steam to the outside of the paper. This necessitates the complete saturation of the several thicknesses of paper and the adhesive between the different layers, and consequently requires a great deal of time, a vast amount of steam and the use of a separate tool or scraper for removing the paper after saturation. The process has other very objectionable features, among these being the large amount of escaping steam filling the apartment and damaging wood-work, paint, furniture and decorations. It also litters the room with scraps of slimy, sticky, water-soaked paper. These and many other objectionable features are done away with in the improved device.

The objects of this invention are to provide a device of simple and economical construction, that is easily and conveniently operated.

Another object of the invention is to provide a device that combines a scraper and moistener, to be used in connection with a low pressure steam supply, that operates between the wall and the layer of paper next thereto, or on the line of adhesion of the paper to the wall, softening up the adhesive substance and at the same time stripping the paper loose in large pieces and thereby saving time and overcoming the objection to littering a room with small scraps of paper, plaster and dust.

The peculiar construction of the hand piece or scraper, which is scraper and moistener combined, requires the use of only a small source of steam supply, as a very few pounds of steam pressure are necessary, the object being to soften the adhesive and not to saturate the paper. The work is done between the wall and the layer of paper next thereto, therefore there is only one clinging substance to contend with. The escaping steam is reduced to a minimum, as it is kept under perfect control by a spring valve lever, and is only brought into use when the adhesion is stubborn.

The best results are obtained, in the improved device by the use of very wet steam, this is accomplished by the expansion of the steam after it enters the flaring portion of the head. By the time that it escapes from the perforations along the line of the blade at the adhesion line, the steam has become condensed to such an extent that very little of it escapes into the room, but it will be absorbed by the adhesive substance and loosen the paper from the wall. The preferable use of wet steam also has advantages in lessening the cost of the boiler, when a portable boiler is used, as a source of steam supply.

In the drawings forming a part of this specification Figure 1 illustrates the device connected with a portable steam generator or boiler. Fig. 2 is a vertical section of the improved device. Fig. 3 is an elevation in direction of arrow 3, Fig. 2, with a portion of the handle broken away. Fig. 4 is a fragmentary view of the reverse side of the head piece. Fig. 5 is a vertical section of a different form of head, used for scraping downward, and Fig. 6 is an elevation in direction of arrow Fig. 5.

2   1,142,811

Similar characters of reference designate similar parts in the drawings and specification.

The improved device is used in connection
5 with a steam supply, preferably a small portable boiler as shown in Fig. 1, to which it is connected by steam hose of sufficient length to admit of convenient manipulation over a considerable wall area in a room,
10 without necessitating the frequent moving of the steam generator. The boiler as shown in the drawings, is no part of the invention, as the device can be connected with any suitable steam supply. In many instances it
15 may be desirable and convenient to connect with some stationary steam apparatus already installed in the apartment where the device is to be operated, and it will be seen that I claim the steam supply source broadly
20 in combination with the elements of the moistener and scraper. The boiler with oil burner is shown as a simple and convenient accessory.

Referring now to the drawings, the core 1,
25 of the hand piece of the device, is tubular and is adapted to be connected by means of the hose 2 with the steam supply 3, thereby allowing steam to pass from the steam supply to the interior chamber 1ᵃ, of the hand
30 piece, when the valve 3ᵃ is open. The core is insulated from the heat by an asbestos jacket 1ᵃ, outside of which is the handle 1ᵇ. The hand piece is provided with a chamber 1ᶜ. A port 1ᵈ connects the chamber 1ᶜ with the
35 main chamber 1ᵃ, which is connected with the hose 2. A valve 4 is provided for opening and closing the port 1ᵈ; this valve is controlled by the spring lever 4ᵃ, fulcrumed to the hand piece at 1ᶠ. The stem of the valve
40 is adapted to work through a suitable stuffing gland 5. The lever is arranged in such a position, relative to the handle 1ᵇ, that it can be conveniently gripped by the operator, thereby assuring perfect control of the steam
45 supply from the chamber 1ᵃ to the chamber 1ᶜ through the port 1ᵈ.

The head 6 is provided with a flange 6ᵃ, adapted to be bolted to the flange 1ᵉ, a gasket 7 being interposed between the said
50 flanges. The head is provided with a chamber 6ᵇ, adapted to register with the chamber 1ᶜ of the hand piece. The chamber 6ᵇ tapers and flares toward the opening 6ᶜ, in juxtaposition to the scraping edge of a blade 8,
55 which is attached to the head 6. The best results are obtained by the use of a flexible blade, for which reason it is preferable to adjustably attach the blade to the head at the points 6ᵈ, more or less remote from the
60 scraping edge, and at an angle that will allow a clearance between the blade and the edge of the head. Serrations or perforations 6ᵉ are cut in the flared edge of the head, and connect with the chamber 6ᵇ. When the

blade 8 is pressed against the wall, it closes 65 the opening 6ᶜ, but the steam is allowed to escape through the serrations 6ᵉ onto the scraping edge of the blade.

In the operation of the device, under frequent conditions, large areas of paper may 70 be removed by using the scraper alone, with the valve 4 closed, but when the adhesive medium becomes stubborn, the opening of the valve 4 allows the steam to enter the chamber 6ᵇ, and to escape therefrom through 75 the perforations or serrations 6ᵉ, along the stripping edge of the blade 8, and soften the adhesive and thereby permit the blade to readily strip the paper clear from the wall.

The head portion of the device may be 80 made in various shapes to meet different conditions; as for example, in scraping downward on the wall, it may be desirable to use a head piece similar to that shown in Figs. 5 and 6, while in stripping upward a head 85 as shown in the other figures is used, the principle being the same in either case.

While I have shown and described herein a practical working device, embodying my invention, I do not wish to be confined to the 90 detailed construction only so far as required by the scope of the claims and the existing state of the art.

What I claim and desire to secure by Letters Patent is— 95

1. In a device for stripping wall paper, the combination of a source of steam supply; an applicator having a chamber therein; steam connection between the source of steam supply and said chamber; means for 100 controlling the supply of steam from the steam source to the chamber in the applicator; a metallic stripping blade adjustably attached to the applicator; and openings from the applicator chamber in juxtaposi- 105 tion to the stripping edge of the blade, the said stripping blade being adapted to coact with these openings to control the supply of steam therefrom.

2. In a device for stripping wall paper, the 110 combination of a source of steam supply, and a moistening and stripping member, said member consisting of a hand piece having a chamber therein, said chamber being connected with the steam supply; a head 115 member attached to the hand piece; a chamber in said head member; a port between the chamber in the head member and the chamber in the hand piece; means for opening and closing said port; a metallic stripping 120 blade attached to the head member; and openings from the chamber in the head member in juxtaposition to the stripping edge of the blade.

3. In a device for stripping wall paper, the 125 combination of a source of steam supply, and a moistening and stripping member, said member consisting of a hand piece having a

1,142,811

chamber therein, said chamber being connected with the steam supply; a head member attached to the hand piece; a flaring chamber in said head member; a port be-
5 tween the chambers in the head member and the hand piece; means for opening and closing said port; a metallic stripping blade attached to the head member; and openings from the chamber in the head member in its flared edge in juxtaposition to the stripping 10 edge of the blade.

LOUIS FRED GRISWOLD.

Witnesses:
FRED C. SCHLATTER,
W. J. MASTERSON.

Patent Application Forms

OMB No. 0651-0011 (12/31/86)

Applicant or Patentee: _____ Attorney's
Serial or Patent No.: _____ Docket No.: _____
Filed or Issued: _____
For: _____

VERIFIED STATEMENT (DECLARATION) CLAIMING SMALL ENTITY
STATUS (37 CFR 1.9 (f) and 1.27 (b)) — INDEPENDENT INVENTOR

As a below named inventor, I hereby declare that I qualify as an independent inventor as defined in 37 CFR 1.9 (c) for purposes of paying reduced fees under section 41 (a) and (b) of Title 35, United States Code, to the Patent and Trademark
Office with regard to the invention entitled _____
described in

[ ] the specification filed herewith
[ ] application serial no. _____ , filed _____ .
[ ] patent no. _____ , issued _____ .

I have not assigned, granted, conveyed or licensed and am under no obligation under contract or law to assign, grant, convey
or license, any rights in the invention to any person who could not be classified as an independent inventor under 37 CFR
1.9 (c) if that person had made the invention, or to any concern which would not qualify as a small business concern under
37 CFR 1.9 (d) or a nonprofit organization under 37 CFR 1.9 (e).

Each person, concern or organization to which I have assigned, granted, conveyed, or licensed or am under an obligation
under contract or law to assign, grant, convey, or license any rights in the invention is listed below:

[ ] no such person, concern, or organization
[ ] persons, concerns or organizations listed below*

*NOTE: Separate verified statements are required from each named person, concern or organiza-
tion having rights to the invention averring to their status as small entities. (37 CFR 1.27)

FULL NAME _____
ADDRESS _____

| [ ] INDIVIDUAL | [ ] SMALL BUSINESS CONCERN | [ ] NONPROFIT ORGANIZATION |

FULL NAME _____
ADDRESS _____

| [ ] INDIVIDUAL | [ ] SMALL BUSINESS CONCERN | [ ] NONPROFIT ORGANIZATION |

FULL NAME _____
ADDRESS _____

| [ ] INDIVIDUAL | [ ] SMALL BUSINESS CONCERN | [ ] NONPROFIT ORGANIZATION |

I acknowledge the duty to file, in this application or patent, notification of any change in status resulting in loss of entitle-
ment to small entity status prior to paying, or at the time of paying, the earliest of the issue fee or any maintenance fee
due after the date on which status as a small entity is no longer appropriate. (37 CFR 1.28 (b))

I hereby declare that all statements made herein of my own knowledge are true and that all statements made on information
and belief are believed to be true; and further that these statements were made with the knowledge that willful false statements
and the like so made are punishable by fine or imprisonment, or both, under section 1001 of Title 18 of the United States
Code, and that such willful false statements may jeopardize the validity of the application, any patent issuing thereon, or
any patent to which this verified statement is directed.

NAME OF INVENTOR _____   NAME OF INVENTOR _____   NAME OF INVENTOR _____

Signature of Inventor _____   Signature of Inventor _____   Signature of Inventor _____

Date _____   Date _____   Date _____

Form PTO-FB-A410 (8-83)

DECLARATION FOR PATENT APPLICATION	Docket Number (Optional)

As a below named inventor, I hereby declare that:

My residence, post office address and citizenship are as stated below next to my name.

I believe I am the original, first and sole inventor (if only one name is listed below) or an original, first and joint inventor (if plural names are listed below) of the subject matter which is claimed and for which a patent is sought on the invention entitled _____ , the specification of which

is attached hereto unless the following box is checked:

☐ was filed on _____ as United States Application Number or PCT International Application
Number _____ and was amended on _____ (if applicable).

I hereby state that I have reviewed and understand the contents of the above identified specification, including the claims, as amended by any amendment referred to above.

I acknowledge the duty to disclose information which is material to patentability as defined in Title 37, Code of Federal Regulations, § 1.56.

I hereby claim foreign priority benefits under Title 35, United States Code, § 119 of any foreign application(s) for patent or inventor's certificate listed below and have also identified below any foreign application for patent or inventor's certificate having a filing date before that of the application on which priority is claimed.

Prior Foreign Application(s)

			Priority Claimed
_____	_____	_____	☐ Yes ☐ No
(Number)	(Country)	(Day/Month/Year Filed)	
_____	_____	_____	☐ Yes ☐ No
(Number)	(Country)	(Day/Month/Year Filed)	
_____	_____	_____	☐ Yes ☐ No
(Number)	(Country)	(Day/Month/Year Filed)	

I hereby claim the benefit under Title 35, United States Code, § 120 of any United States application(s) listed below and, insofar as the subject matter of each of the claims of this application is not disclosed in the prior United States application in the manner provided by the first paragraph of Title 35, United States Code, § 112, I acknowledge the duty to disclose information which is material to patentability as defined in Title 37, Code of Federal Regulations, § 1.56 which became available between the filing date of the prior application and the national or PCT international filing date of this application.

_____	_____	_____
(Application Number)	(Filing Date)	(Status — patented, pending, abandoned)
_____	_____	_____
(Application Number)	(Filing Date)	(Status — patented, pending, abandoned)

I hereby appoint the following attorney(s) and/or agent(s) to prosecute this application and to transact all business in the Patent and Trademark Office connected therewith:

_____

Address all telephone calls to _____ at telephone number _____
Address all correspondence to _____
_____
_____
_____

I hereby declare that all statements made herein of my own knowledge are true and that all statements made on information and belief are believed to be true; and further that these statements were made with the knowledge that willful false statements and the like so made are punishable by fine or imprisonment, or both, under Section 1001 of Title 18 of the United States Code and that such willful false statements may jeopardize the validity of the application or any patent issued thereon.

Full name of sole or first inventor (given name, family name) _____
Inventor's signature _____ Date _____
Residence _____ Citizenship _____
Post Office Address _____

Full name of second joint inventor, if any (given name, family name) _____
Second Inventor's signature _____ Date _____
Residence _____ Citizenship _____
Post Office Address _____

☐ Additional inventors are being named on separately numbered sheets attached hereto.

Patent and Trademark Office: U.S. DEPARTMENT OF COMMERCE

PATENT APPLICATION TRANSMITTAL LETTER	Docket Number (Optional)

To the Commissioner of Patents and Trademarks:

Transmitted herewith for filing under 35 U.S.C. 111 and 37 CFR 1.53 is the patent application of

_____

entitled _____

_____

Enclosed are:

☐ _____ pages of written description, claims and abstract.

☐ _____ sheets of drawings.

☐ an assignment of the invention to _____

☐ executed declaration of the inventors.

☐ a certified copy of a _____ application.

☐ associate power of attorney.

☐ a verified statement to establish small entity status under 37 CFR 1.9 and 1.27.

☐ information disclosure statement

☐ preliminary amendment

☐ other: _____ .

CLAIMS AS FILED

	NUMBER FILED	NUMBER EXTRA	RATE	FEE
BASIC FEE			$690	$690
TOTAL CLAIMS	- 20 =	*	x $20	
INDEPENDENT CLAIMS	- 3 =	*	x $72	
MULTIPLE DEPENDENT CLAIM PRESENT			$220	

* NUMBER EXTRA MUST BE ZERO OR LARGER

TOTAL $ _____

If applicant has small entity status under 37 CFR 1.9 and 1.27, then divide total fee by 2, and enter amount here. **SMALL ENTITY TOTAL** $ _____

☐ A check in the amount of $ _____ to cover the filing fee is enclosed.

☐ The Commissioner is hereby authorized to charge and credit Deposit Account No. _____ as described below. I have enclosed a duplicate copy of this sheet.

  ☐ Charge the amount of $ _____ as filing fee.

  ☐ Credit any overpayment.

  ☐ Charge any additional filing fees required under 37 CFR 1.16 and 1.17.

  ☐ Charge the issue fee set in 37 CFR 1.18 at the mailing of the Notice of Allowance, pursuant to 37 CFR 1.311(b).

_____   _____
Date   Signature

                                 _____
                                 Typed or printed name

                                 _____
                                 Address

Patent and Trademark Office: U.S. DEPARTMENT OF COMMERCE

AMENDMENT TRANSMITTAL LETTER		Docket Number (Optional)

Application Number	Filing Date	Examiner	Group Art Unit

Invention Title

TO THE COMMISSIONER OF PATENTS AND TRADEMARKS

Transmitted herewith is an amendment in the above - identified application.

☐ Small Entity status of this application has been established under 37 CFR 1.27 by a verified statement previously submitted.

☐ A verified statement to establish Small Entity status under 37 CFR 1.27 is enclosed.

☐ No additional fee is required.

☐ The fee has been calculated as shown below:

CLAIMS AS AMENDED

	(1) CLAIMS REMAINING AFTER AMENDMENT	(2) HIGHEST NUMBER PREVIOUSLY PAID FOR	(3) PRESENT NUMBER EXTRA	RATE	FEE
TOTAL CLAIMS	*	minus **		x $20	
INDEPENDENT CLAIMS	*	minus ***		x $72	
MULTIPLE DEPENDENT CLAIM ADDED				$220	
				TOTAL	$
If applicant has small entity status under 37 CFR 1.9 and 1.27, then divide total fee by 2, and enter amount here.			SMALL ENTITY TOTAL		$

* If the entry in column 1 is less than the entry in column 2, write "0" in column 3

** If the highest number previously paid for IN THIS SPACE is less than 20, enter "20".

*** If the highest number previously paid for IN THIS SPACE is less than 3, enter "3".

The "highest number previously paid for" (total or independent) is the highest number found in the appropriate box in column 1.

☐ Please charge Deposit Account Number _____ in the amount of $ _____ .
A duplicate copy of this sheet is enclosed.

☐ A check in the amount of $ _____ to cover the filing fee is enclosed.

☐ The Commissioner is hereby authorized to charge payment of the following fees associated with this communication or credit any overpayment to Deposit Account Number _____ .
A duplicate copy of this sheet is enclosed.

☐ Any additional filing fees required under 37 CFR 1.16.

☐ Any patent application processing fees under 37 CFR 1.17.

_____	_____
Date	Signature

Patent and Trademark Office: U.S. DEPARTMENT OF COMMERCE

Patent Appendix VIII

A Sample Agent Agreement

REPRESENTATION AGREEMENT

REPRESENTATION AGREEMENT, made on _____199___., between, Gary Ahlert, D.B.A., The Creative Group, 400 Main Street, Stamford, CT 06901 (hereinafter referred to as TCG) and

This Agreement consists of _____( ) typewritten pages including Schedule A.

WHEREAS, TCG is engaged in the business of representing artistic and scientific properties for the purposes of securing revenues from persons who will be licensed to manufacture and sell articles relating to such properties, and

WHEREAS, the OWNER is the inventor of properties suitable for licensing.

NOW, THEREFORE, in consideration of the premises and of the mutual promises and agreements thereinafter set forth, the parties hereto agree as follows:

1.   REPRESENTATION
        OWNER hereby appoints TCG as the sole and exclusive representative for licensing, sale or merchandising of the property listed on attached Schedule A, annexed hereto and made part of this agreement. TCG will have the exclusive right to grant licenses for use, manufacture, distribution, sale, advertising and promotion throughout the world on the property listed in Schedule A and any trademarks related to this property. The OWNER agrees to refer to TCG all inquiries relating to the licensing or merchandising rights with respect to the property set forth in Schedule A.
        a. TCG shall have the right of first refusal on any additional properties created by OWNER.
        b. TCG will advise OWNER within thirty (30) days of receipt of such properties, of it's acceptance or rejection of such properties.
        c. If accepted by TCG, these properties will be added to Schedule A and made part of this agreement.
2.   DUTIES OF TCG
        TCG will use its best efforts to promote, sell, license, or merchandise the OWNER'S property listed in Schedule A so as to obtain the largest gross receipts from such sale, licensing, merchandising, or any other Agreements for the benefit of the OWNER in regard to the property listed in Schedule A.

3.   ROYALTY PAYMENTS
        TCG and OWNER agree to the following terms:
        (a)   Any monies paid as a cash advance or "up front" fees will be divided as follows: fifty (50%) percent to TCG and fifty (50%) percent to OWNER.

        (b)   Annual net royalties from any license, merchandising or other Agreement will be paid in the following manner: fifty (50%) percent to TCG and fifty (50%) percent to OWNER.

94

-2-

4. SALE

In the event TCG negotiates the sale of property listed in Schedule A, TCG will be paid fifty (50%) percent of the gross sales price.

5. COMPENSATION

(a)   All royalty payments will be made directly to TCG.

(b)   Within thirty (30) days after the receipt of sales statement and/or applicable payment, TCG will prepare and forward to OWNER, a statement indicating the amount of sales, together with any royalty payment, if due, minus necessary costs expended in the performance of this Agreement by TCG.

(c)   Necessary costs are defined as telephone, postage, travel, online research services and attorneys fees.

(d)   TCG agrees to keep accurate books of accounts and records covering all transactions relating to this Agreement, and OWNER'S duly authorized representatives will have the right at any time during reasonable business hours, to examine such records insofar as they relate to the subject matter of this Agreement.

6. LICENSES, SALES, MERCHANDISING AND OTHER AGREEMENTS

(a)   All license, sale, merchandising or any other Agreement made on OWNER'S behalf, including, sub licenses, or any amendments or modification thereto negotiated by TCG with third parties in accordance with the terms of this Agreement will be subject to the prior approval of the OWNER. Such approval will not be unreasonably withheld and OWNER will signify its approval or disapproval within fifteen (15) days from the submission of said license or other Agreement. If licensing or other Agreement are entered into with companies in foreign markets which require sub licensing Agreements, the OWNER authorizes the foreign licensee or contractor to be responsible for entering into all such sub-licensing Agreements.

7. TERM OF REPRESENTATION FOR NEGOTIATION.

(a)   The term of this Agreement will commence on the date hereof and will expire (1) year from the date hereof on,

(b)   If a license, merchandising or other Agreement is obtained or negotiation is initiated within the aforementioned one (1) year period, the term of this Agreement will be automatically extended for a two (2) year period commencing on

Thereafter the term of this Agreement will automatically be extended from year to year unless terminated by either party by written notice to the other at least sixty (60) days prior to the anniversary date mentioned in paragraph 7(a) and 7(b) of this Agreement.

(c)   If TCG negotiates a license or other Agreement on behalf of OWNER in regard to property listed in Schedule A, TCG will receive royalties in accordance with paragraph 3(a) and 3(b) of this Agreement, for the term of such license or other Agreement or any renewals thereof.

-3-

(d)   Upon termination of this Representation Agreement, pursuant to paragraphs 7(a) and 7(b) of this Agreement, the OWNER may not enter into any new licensing agreements for a period of three (3) years with any of the then existing licensees wherein TCG was deemed a party to the licensing agreements entered into during the duration of this Representation Agreement unless provision is made to guarantee that any income derived from such an agreement will be divided in accordance with paragraphs 3(a) and 3(b) of this Agreement.

(e)   If either party will cancel this Agreement pursuant to paragraphs 7(a) and 7(b), TCG will have the right to conclude Agreements currently being negotiated by TCG for a period of up to six (6) months after the effective date of such termination,, TCG must notify the OWNER in writing within fifteen (15) days after the effective date of such termination advising the OWNER of such parties with whom TCG is in negotiation. Any Agreements signed within this six (6) month extension period will be considered as if they were signed during the term of the Representation Agreement and will continue to be serviced by TCG during the term of such Agreement and any extension and/or renewals thereof.

(f)   This Agreement will survive the death or disability of the parties.

8.   OWNER'S RIGHTS AND RESPONSIBILITIES IN REGARD TO LITIGATION

(a)   The OWNER represents and warrants that the OWNER has the rights and the authority to grant the rights herein granted to TCG and that the use of Property as herein contemplated, to the best of OWNER'S knowledge, will not infringe upon the rights of any other person or corporation. Additionally, OWNER agrees to not transfer or assign any of the aforementioned rights, including patents, trademarks, or copyrights to any other party unless mutually agreed to in writing and signed by both parties. If litigation arises as a result of the OWNER'S misrepresentation or wrongdoing in regard to the product listed in Schedule A, OWNER agrees to indemnify TCG for any liabilities arising from such misrepresentation or wrongdoing.

(b)   The OWNER agrees to maintain and defend the OWNER'S rights in the property which are the subject of any License, sale, merchandising or other agreement hereunder. The OWNER has the right to determine the extent to which the OWNER will undertake to defend Licensees and TCG from claims or liabilities arising out of or in connection with use of the Property granted by the OWNER as herein contemplated.

9.   RELATIONSHIP OF PARTIES

This Agreement does not constitute and will not be construed as constituting a partnership or joint venture between TCG and the OWNER, and neither party will have any right to obligate or bind the other in any manner whatsoever except as authorized in this Agreement, and nothing therein contained will give or is intended to give any rights of any kind to any third parties.

10.   MISCELLANEOUS

(a)   This Agreement will be construed in accordance with the laws of the State of Connecticut.

(b)   All notices will be in writing.

(c)   This Agreement can only be changed by an Agreement in writing signed bothparties.

-4-

(d)   This Agreement may not be transferred without the written approval of both parties; however, in the event that this Agreement is terminated due to any reason gross receipts payable under the terms of existing Licensing Agreements and/or renewals, will be paid to the parties involved or their designated beneficiaries.

IN WITNESS WHEREOF, the parties hereto have executed this Agreement as of the day and year first written above.

ATTEST:

_____ _____

Gary Ahlert, DBA The Creative Group

ATTEST:

_____

-5-

SCHEDULE A

Should any enhancements, modifications or additional patents be required for the further development of the property listed on this Schedule A, they too will become part of this Schedule A and this agreement.

PROPERTY:

Patent Appendix IX

A Sample Licensing Agreement

LICENSE AGREEMENT

LICENSE AGREEMENT made ____, by and between Gary Ahlert, D.B.A. The Creative Group, a Connecticut business, located at 400 Main Street, Stamford, Connecticut 06901 (hereinafter referred to as "Agent") and _____ (hereinafter referred to as "Licensee", and _____, located at , (hereinafter referred to as "Licensor").

The specific details of this License to which the language in succeeding numbered paragraphs hereto refer, and which said language embellishes and explains, is set forth in the immediately following Schedules A through G: The headings noted as defined in the body of the license form.

Schedule A: PROPERTY:

Schedule B: LICENSED PRODUCTS:

Schedule C: LICENSED TERRITORY:

Schedule D: LICENSED PERIOD:

                     OPTION TO RENEW:

                     Licensee has an option to renew for additional
                     two (2) one year periods if Licensee guarantees sales of
                     $_____ annually or guarantees to make a percentage
                     compensation   payment   equal   to   the guaranteed
                     sales. Licensee must notify Licensor 90 days prior
                     to the expiration of the initial term of this Agreement
                     of its intention to renew or 90 days prior to the expiration
                     of any renewal term of this Agreement.

                     Automatically renewed thereafter for twelve
                     (12) month periods, unless either party gives notice of
                     its desire not to renew 90 days prior to the expiration
                     of the initial term of this Agreement, or 90 days prior
                     to the expiration of any renewal term of this Agreement.

Schedule E: PERCENTAGE COMPENSATION:

-2

GUARANTEE:

Schedule F: COPYRIGHT, PATENT & TRADEMARK NOTICES:

                   Copyright Notice:
                   Trademark Notice:
                   Patent Notice:

Schedule G: MARKETING DATE:

THIS WILL CONFIRM OUR AGREEMENT AS FOLLOWS:

1.   GRANT OF LICENSE: Licensor grants to Licensee for the term of this Agreement subject to the terms and conditions hereinafter contained, the license to utilize the property and trademark described in Schedule A, as shown above (herein such property and trademark are collectively called "Property & Trademark").   The Licensee shall use the license granted hereunder solely in connection with the manufacture, distribution and sale of the article or articles specified in Schedule B as shown above (herein such article or articles are called "Licensed Products").

2.   TERRITORY: The Licensee shall be entitled to use the License granted hereunder only in the territory described in Schedule C as shown above (herein such territory is called "Licensed Territory"). Licensee will not make or authorize any use of the License or Licensed Products outside the Licensed Territory.

3.   LICENSE PERIOD: The license granted hereunder shall be effective and terminate as of the dates specified in Schedule D, as shown above, unless soon terminated or renewed in accordance with the terms and conditions hereof.

4.   PAYMENT: A. Percentage Compensation: Licensee agrees to pay Agent, acting as Licensor, a sum equal to the percentage specified in Schedule E of all net sales by Licensee or any of its affiliated, associated or subsidiary companies of the Licensed Products covered by this Agreement.   (Such percentage of net sales is herein called "Percentage Compensation").   Said payments shall be remitted by Licensee concurrently with each of the periodic statements required below. The term "net sales" shall mean gross sales less quantity discounts and returns, but no deduction shall be made for uncollectable accounts.   No costs incurred in the manufacture, sale, distribution exploitation of the Licensed Products shall be deducted from any Percentage Compensation payable by Licensee.   Said Percentage Compensation shall also be paid by Licensee to Agent, acting for Licensor, on all Licensed Products

-3-

distributed by Licensee or any of its affiliated, associated or subsidiary companies even if not billed, such as free introductory offers, samples, etc., and shall be based upon the usual net sales price for such Licensed Products sold to the trade by Licensee.

B. All payments to Licensor should be made to and remitted to The Creative Group acting as Agent for Licensor.

5.   PERIODIC STATEMENTS: Within thirty (30) days after the initial shipment of the Licensed Products covered by this Agreement, and promptly on the 30th day of every Jan., April, July and Oct. thereafter, Licensee shall furnish to Agent complete and accurate statements, certified to be accurate by Licensee, or if a corporation, by an officer of Licensee, showing the number, description and gross sales price, and net sales price of the Licensed Products distributed and/or sold by Licensee during the preceding calendar quarter. Such statements shall be furnished to Agent whether or not any of the Licensed Products have been sold during the calendar quarter in which such statements are due. Receipt or acceptance by Agent of any of the statements furnished pursuant to this Agreement or of any sums paid hereunder shall not preclude Licensor or Agent from questioning the correctness thereof at any time, and in the event that any inconsistencies or mistakes are discovered in such statements or payments, they shall immediately be rectified and the appropriate payments made by Licensee. Upon demand of Licensor or Agent, Licensee shall, at its own expense, but not more than one (1) time in any twelve (12) month period, furnished to Licensor a detailed statement by an independent certified public accountant showing the number, description, gross sales price, itemized deductions from gross sales price and net sales price of the Licensed Products covered by this Agreement distributed and/or sold by Licensee to the date of Licensor's demand.

6.   BOOKS AND RECORDS: Licensee agrees to keep accurate books of accounts and records covering all transactions relating to the license hereby granted, and Licensor or Agent and its duly authorized representatives shall have the right at all reasonable hours of the day to an examination of said books of account and records and of all other documents and material in the possession or under the control of Licensee with respect to the subject matter and the terms of this Agreement, and shall have free and full access, thereto, for said purposes and for the purpose of making extracts therefrom. Licensor and Agent agree that they will conduct no more than one (1) examination pursuant to the terms of this Article during any twelve (12) month period that this Agreement is in effect. All books of account and records shall be kept available for at least two (2) years after the termination of this license, or any renewal thereof, and Licensee agrees to permit inspection thereof by Licensor during such two (2) year period.

7.   INDEMNIFICATIONS: Licensee hereby indemnifies and agrees to hold Licensor and Agent harmless from any claim or suits arising out of any unauthorized use of any process method or device by Licensee in connection with Licensed Products covered by this Agreement and from any claim or suits arising out of Licensee's manufacture and sale of said Licensed Products. Licensee agrees to obtain, at its own cost and expense, product liability insurance providing adequate protection for Licensor and Agent against any such claims or suits arising from any alleged defects on said Licensed Product, and within thirty (30) days from the date hereof, Licensee will submit to Licensor and Agent proof of a fully paid policy of insurance, naming Licensor and Agent as additional insured parties.

-4-

8.   COPYRIGHT, PATENT, AND TRADEMARK NOTICES: The Licensee shall cause to be imprinted irremovable and legibly on each Licensed Product manufactured, distributed or sold by the Licensee under this Agreement a copyright notice including the letter "c" enclosed in a circle, all rights reserved, (ADD YEAR AND DATE) ____. Additional wording, such as "produced under license from _____ is permissible, but in no event will the foregoing copyright notice be eliminated or modified in the absence of written approval from. Where it is impractical to have the full copyright notice appear on the products as set forth, Licensee may use the copyright c with the initials   . Where such abbreviated notice is used, the full name described in Schedule F, shall be legibly imprinted on some accessible portion of the Licensed Products. Licensee also agrees to cause the appropriate statutory notice of trademark registration to be imprinted wherever the Property trademark is used. In the event the Licensed Product is marketed in any container bearing the Property, the aforesaid copyright notice and/or trademark registration notice shall also appear on said container. Licensee agrees to deliver to Licensor ten (10) of the Licensed Products upon which the Property and Design is imprinted for copyright registration and/or trademark registration in the name of the person, firm or corporation described in Schedule F, as shown above, in compliance with Federal laws relating to copyrights and trademark.

(a)   In addition to copyright/trademark notices, appropriate patent identification must be imprinted on each licensed product as described above and in accordance with Patent and Trademark regulations.

9.   APPROVALS: The Licensee agrees to furnish Licensor, free of cost, for its written approval as to quality and style, samples of the Licensed Products covered by this Agreement before their manufacture or sale, whichever first occurs, and no Licensed Product shall be manufactured or sold by the Licensee without such prior written approval. If the Licensed Product is changed significantly from the approved samples, then new samples must be submitted for Licensor's written approval. Subject, in each instance, to the prior written approval of Licensor, the Licensee or its agents may use textual and/or pictorial matter pertaining to the property on such promotional display and advertising material as may, in its judgment, promote the sale of the Licensed Products. Approval by the Licensor will not be unreasonably withheld.

10.   DISTRIBUTION: The Licensee shall sell Licensed Products either wholesales, to jobbers, or distributors for resale and distribution to retail stores and merchants for resale and distribution directly to the public. In the event the Licensee sells or distributes a Licensed Product at a special price directly or indirectly to itself, including, without limitation by specification, any subsidiary of the Licensee, or to any other person, firm or corporation related in any manner to the Licensee or its officers. directors or major stockholders, the Licensee shall pay compensation with respect to such sales or distribution based upon the price generally charged the trade by Licensee.

11.   GOOD WILL: The Licensee acknowledges that good will associated with the Property exclusively belongs to Licensor and its grantors and that any trademark usage acquiring a secondary meaning in the mind of the purchasing public belongs to the Licensor.

12.   SPECIFIC UNDERTAKINGS OF LICENSEE: During the License Period each additional License Period and thereafter, the Licensee agrees that:

(a)   It will not attacked the title of Licensor or its grantors in and to the Property or any copyright or trademark pertaining thereto, nor will it attack the validity of the License granted hereunder;

(b)   It will not harm, misuse or bring into disrepute the Property;

(c) It will manufacture, sell and distribute the Licensed Products in an ethical manner and in accordance with the terms and intent of this Agreement;

(d) It will not create any expenses chargeable to Licensor and Agent without the prior written approval of Licensor and Agent;

(e) It will protect to the best of its ability its right to manufacture, sell and distribute the Licensed Products hereunder;

(f) It will not, without the prior written consent of Licensor and Agent enter into any sublicense or agency agreement for the manufacture, sale, or distribution of the Licensed Products; and

(g) It will not enter into an agreement relating to the Property and Designs for commercial tie-ups or promotions with any company engaged in whole or in part in the production of motion pictures and television without the prior written consent of Licensor and Agent.

13.   DEFAULT, BANKRUPTCY, VIOLATION, ETC.:

A. In the event the Licensee does not commence in good faith to manufacture, distribute and sell each Licensed Product in substantial quantities on or before the date specified in Schedule G, as shown above, Licensor, in addition to all other remedies available to it, shall have the option to terminate the license granted hereunder with respect to such Licensed Product upon mailing to the Licensee notice of such termination.

B. In the event the Licensee files a petition in bankruptcy or if a petition in bankruptcy is filed against the Licensee or if the Licensee becomes insolvent or makes an assignment for the benefit of its creditors or any arrangement pursuant to any bankruptcy law or if Licensee discontinues its business or if a receiver is appointed for it or its business, the license granted hereunder, without notice, shall terminate automatically upon the occurrence of any such event. In the event the license granted hereunder is so terminated neither the Licensee nor its receiver, representatives, trustees, agents, administrators, successors and/or assigns shall have any right to sell, exploit or in any way deal with or in any Licensed Product or any carton, packing or wrapping material, advertising, promotional or display material pertaining to any Licensed Product.

C. If Licensee shall violate any of its other obligations under the terms of this Agreement, Licensor shall have the right to terminate the license hereby granted upon thirty (30) days' notice in writing, and such notice or termination shall become effective unless Licensee shall completely remedy the violation within the thirty-day period and satisfy Licensor that such violation has been remedied.

D. In the event the License granted hereunder is terminated in accordance with the provision of Article 13, subdivision C, hereof, all compensation theretofore accrued, plus any additional Advance Compensation not yet paid and Guaranteed Compensation not yet paid, shall become due and payable immediately to Agent, acting for Licensor, who shall not be obligated to reimburse the Licensee for any Advance Compensation or additional Advance Compensation.

14.   FINAL STATEMENT UPON TERMINATION OR EXPIRATION:

A. Licensee shall deliver, as soon as practicable, to Licensor, a statement indicating the number and description of Licensed Products on hand or in process of manufacture as of: (a) sixty (60) days prior to the expiration of the Licensed Period and each additional Licensed Period; and (b) fifteen (15) days after receipt from Licensor of a notice terminating the license granted hereunder or, in the event no such notice is required, fifteen (15) days after the occurrence of any event which terminates such license. Licensor of the Agent shall have the option to conduct a physical inventory in order to ascertain or verify such

-6-

inventory and/or statement. In the event the Licensee refuses to permit Licensor or Agent to conduct suchphysical inventory, the Licensee shall forfeit its rights hereunder to dispose of such inventory. In addition to such forfeiture, Licensor and Agent shall have recourse to all other remedies available to them.

B. If this Agreement is terminated in accordance with the expiration of the License Period, Licensee shall have a period of six (6) months from the date of termination thereof to sell its existing supply of Licensed Products provided that upon termination of this Agreement, Licensee returns all art work and reproductions thereof to Licensor, destroys all prototypes made by it from Licensor's art work and reproductions, and discontinues all further manufacture of Licensed Products and/or packaging theretofore manufactured and sold by it under this Agreement.

15.   NOTICES: All notices and statements provided for herein shall be in writing and together with all payments provided for herein shall be mailed to the addresses set forth above or such other address as may be designated in writing by Licensor and Agent or Licensee from time to time.

16.   NO PARTNERSHIP, ETC.: This Agreement does not constitute and shall not be construed as constituting a partnership or joint venture between Licensor, Agent and the Licensee. The Licensee shall have no right to obligate or bind Licensor and Agent in any manner whatsoever, and nothing herein contained shall give or is intended to give any rights of any kind to any third persons.

17.   NON-ASSIGNABILITY: This Agreement shall bind and inure to the benefit of Licensor and Agent its successors and assigns, shall bind the Licensee, its successors and assigns, but shall not be assignable by the Licensee and shall inure to the benefit only of the Licensee but not its successors or assigns.

18.   CONSTRUCTION: This Agreement shall be construed in accordance with the laws of the State of Connecticut.

IN WITNESS WHEREOF, the parties hereto have signed this Agreement as of the day and year above written.

Attest:

_____ BY: _____

Gary Ahlert, President
THE CREATIVE GROUP

Attest:

_____ BY: _____

Attest:

_____ BY: _____

Patent Appendix X
A Sample Nondisclosure Agreement

THE CREATIVE GROUP

00 Main Street • Suite 210 • Stamford, Ct. • 06901 • U.S.A. • (203) 359-3500

FILE # _____

DECLARATION OF CONFIDENTIALITY
AND
NON-USE

t is agreed that _____
name

esiding at _____
address

_____ _____ _____ _____
city   state   zip   country

nd whose telephone numbers are : Home (   ) _____ Business (   ) _____

nd The Creative Group, its successors and / or assigns, its employees, officers, affiliated consultants, nd agents, will respect the confidentiality of all concepts / products, marks, or copyrightable materials ubmitted and agree that your concept / product, mark, or copyrightable material shall not be used, sold, ssigned or disclosed to any other person, organization, or corporation without your prior approval and ritten permission.

lease sign and date your acknowledgement of this agreement as indicated below.

igned this _____ day of _____ 19 _____

cknowledged by : _____ By : _____

itle : _____

105

THE CREATIVE GROUP

400 Main Street • Suite 210 • Stamford, Conn • 06901 • (203) 359-3500
Fax: (203) 978-1919 • 1-800-678-5306

CONNECTICUT   AGREEMENT OF NON-DISCLOSURE
                      AND CONFIDENTIALITY

THIS AGREEMENT, made this_____ day of_____, 199__ by and between THE CREATIVE GROUP,   400   Main   Street,   Stamford,   CT.,   06901   (hereinafter   referred   to   as   TCG) and_____.

WHEREAS, TCG is in the business of representing inventors and individuals as agents for the purpose of licensing, marketing and sales of their products.
.PRODUCT:_____
_____

WHEREAS, TCG is desirous of disclosing such ideas and products to _____for the purpose of developing, and/or manufacturing, and/or marketing certain products and both parties desire and agree that all information remain confidential with the parties involved.

WHEREAS, TCG is desirous of protecting proprietary secrets, processes, procedures, sources and market position from unfair competition.

WHEREAS, _____is desirous of working with TCG in the development and/or manufacturing, and/or marketing and sale of certain products and services both parties agree that all information remain confidential with the parties involved.

NOW THEREFORE, it is agreed as follows:

1.   The undersigned hereby mutually agree that all confidential information whether printed, written or oral, furnished by one party, its agents, successors, employees or assigns shall be held in confidence by the other undersigned party, its agents, employees, successors or assigns and shall not divulge to any third party, in any manner whatsoever, directly or indirectly, except with the expressed written permission of the other party.

2.   The parties hereto agree and understand that during the course of their performance of the Agreement, they will become privy to certain confidential information relating to the business of the other party. As used herein, confidential information means any information including, without imitation, any patents, processes, discoveries, manufacturers, customers, suppliers, resources, marketing plans or compilation of information or names which (a) is used in the business of the other party of results from its research and marketing activities (b) is private of confidential in that it is not generally known or available to the public and (c) gives the other party an opportunity to obtain an advantage over the other party

-1-

AGREEMENT OF NON-DISCLOSURE

_____ agrees that it shall not, for a period of three (3) years, use or disclose any such confidential information or use such confidential information to compete, directly or indirectly, with TCG and agrees that it will not, at any time, call upon any manufacturer, supplier or customer of TCG for the purpose of soliciting the manufacture and the sale of products described herein and will not otherwise interfere with the business relationship of TCG with it suppliers, customers or clients.

     3.   Disclosure of any of the confidential information provided by TCG to_____ to any third party would constitute a breach of this Agreement and will entitle TCG, its agents, successors and assigns to such actions at law of equity for damages and/or injunctive relief.

The undersigned hereby acknowledge that the disclosure of any confidential information to the undersigned by the other party is sufficient consideration to support this Agreement.

This Agreement will be governed by the laws of the State of Connecticut.

In Witness, Whereof, the undersigned have caused this document to be signed this the_____day of_____, 19___.

BY:_____Title:_____

BY:_____Title:_____

THE CREATIVE GROUP

BY:_____
    Gary Ahlert, President

PART

Copyrights

CHAPTER **6**

Copyright Basics

In this section we'll explore a different sort of intellectual property generally grouped under the somewhat vague heading "artistic expression." We'll examine how to protect your artistic expressions through action based on an understanding of the copyright laws. We'll discuss what a copyright is and what it protects, and how to both register one and transfer it to others. We'll also look into some basic methods of marketing and licensing your artistic expressions—and what to do if someone infringes on your copyright.

What Is a Copyright?

A *copyright* is a form of intellectual property protection provided by the law of the United States (Title 17, U.S. Code) to the authors of artistic expressions. While the concept of copyright protection encompasses work obviously related to the fine arts (for example, painting, writing, music composition, and sculpture), it also includes other forms of expression no less artistic but perhaps less obvious (architectural drawings, computer programs, scientific notebooks, and the like).

The best way to understand the copyright is *as* a right. Citizens of the United States do not have to apply for freedom of speech; they automatically possess it by virtue of their citizenship. Likewise, the act of creating something in a *tangible* form (e.g., writing it down or painting it) engenders certain intellectual property rights. If you are the creator of such an original work, your work is automatically copyrighted at the moment of its creation, and you are automatically entitled to any applicable benefits. A work is considered *created* when it is fixed in a *copy* or a *phonorecord* for the first time.

Copies or duplicates are material objects from which a work can be read or visually perceived either directly or with the aid of a machine or device. These include books, manuscripts, sheet music, film, videotape, and microfilm. *Phonorecords* are material objects embodying fixations of sounds (excluding, by statutory definition, motion-picture soundtracks). This category includes audio tapes and phonograph disks. Thus, for example, a song (the "work") can be fixed in sheet music ("copies") or in a compact disc ("phonorecords"), or both.

If a work is prepared over a period of time, the part of the work that is fixed on a particular date constitutes the created work as of that date.

History

For almost a thousand years, since the beginnings of English Common Law in the year 1066, the works of authors and painters have been recognized as their property. In 1710 this notion was formalized in England, in the Statute of Anne, which specifically protected authors against plagiarism. That concept of protection forms the basis for copyright law to this day.

In the United States, copyright law derives its authority from the same source as does patent law—Article I, Section 8, of the Constitution, which authorizes Congress to enact laws protecting the rights of inventors and authors. In 1790, soon after the ratification of the Constitution, Congress used this power to pass the first copyright laws, which (in spite of the Revolutionary War) were based directly on the aforementioned Statute of Anne. These original laws covered only books, maps, and charts. Over the years, many other forms of expression were added to what continues to be a growing list. In 1802, prints were added; in 1831, musical compositions; and in 1856, dramas. The banner year 1870 saw the laws extended to paintings, drawings, sculptures, and other forms of fine art. In that year the copyright facilities were also centralized in the Library of Congress.

A 1909 law further expanded the realm of what could be copyrighted. Although it remained in effect for sixty-nine years, the 1909 copyright law was very different in some respects from the law today. For example, the ownership of a federal copyright was dependent on publication or public performance of works, and authors' rights were only secure once the works had been officially registered. The status of unpublished creations was left in the domain of the states. Any protection afforded to authors for work they did not publish varied greatly, depending on where they lived.

The 1976 Copyright Act, the last major revision to date of the United States Copyright Law, came into effect in 1978. Because that change was relatively recent, many people who have been peripherally involved in copyrighting for years are sometimes confused about the new law, and as a result incorrectly or unnecessarily protect their own work, and give faulty advice to others. The fact that, in some instances, different rules apply for works copyrighted prior to 1978 doesn't help matters, either. In any event, please be sure to check the information in this segment of this book before proceeding on someone's well-meant advice.

Although we'll be exploring the details of the new law throughout the remainder of this discussion, it is important to note in this overview that the 1978 Copyright Law in fact did establish a single federal system that protects works from the moment of their creation in a tangible form. It also genuinely created a single term for the copyright, lasting the author's lifetime plus fifty years. And, it included an important new right (which we'll examine later) that causes any transfer of the copyright to be rescindable by the author after thirty-five years.

The Five Rights

The rights granted by the new copyright law can be broken down into five basic categories, as follows. The owner of a copyright has the sole and exclusive right:

- *To reproduce* the copyrighted work in copies or phonorecords
- *To prepare derivative works* based on the copyrighted work
- *To distribute copies or phonorecords* of the copyrighted work to the public by sale or other transfer of ownership, or by rental, lease, or lending
- *To perform the copyrighted work publicly,* in the case of literary, musical, dramatic, and choreographic works, pantomimes, and motion pictures and other audiovisual work
- *To display the copyrighted work publicly,* in the case of literary, musical, dramatic, and choreographic works, pantomimes, and pictorial, graphic, or sculptural works, including the individual images of a motion picture or other audiovisual work

This, in a nutshell, is the copyright law. Keep these five rights in mind as we proceed—we'll be referring back to them in detail. For now, take note that while it is illegal for anyone to violate any of these rights, they are *not* unlimited in scope.

What *Can* Be Copyrighted?

As our short history revealed, the list of copyrightable material has been slowly growing. While the basic notion of an artistic expression has been maintained throughout, the 1976 act fixed into the following nine categories the types of expression that fall under copyright protection:

1. Literary works
2. Musical works, including any accompanying words
3. Dramatic works, including any accompanying music
4. Pantomimes and choreographic works
5. Pictorial, graphic, and sculptural works
6. Motion pictures and other audiovisual works
7. Sound recordings
8. Compilations
9. Derivative works

These categories should be viewed very broadly: Many things you might not necessarily consider as belonging to a particular category *are* legally included. Computer programs, for example, which to the layman are little more than an indecipherable series of formulas and commands, fall under the category "literary works." More clearly, though not necessarily more obviously, maps and architectural plans fall under the category of "pictorial, graphic, and sculptural works."

Sufficient Creative Effort

In an effort to clarify and codify copyright protection, the 1976 act also introduced the following three criteria which must be met in order for a work to be copyrightable:

1. It must fall within one of the nine categories listed above.
2. It must be in a "fixed and tangible medium of expression."
3. It must be "an original work of authorship."

While the first criterion is fairly obvious, let's take a closer look at the second and third.

A Fixed and Tangible Expression

Just as in patent law, copyright extends only to something that has been established or expressed in a certain medium (print, clay, canvas, audio tape, etc.). The embodiment of the expression has to exist only long enough to permit it to be seen, reproduced, or otherwise communicated (by means of a machine, for example) for a moment.

It may surprise you to learn that some expressions which could easily be made fixed and tangible are sometimes not made so, and as such lose copyright protection. Acting improvisations, live news reports, sporting events, and concerts, for example, do *not* meet the "tangible expression" criterion, so don't fall under copyright protection unless they have been somehow recorded.

Original Works of Authorship

The final criterion hinges on the use and meaning of the word *original.* To qualify for copyright protection, a work has to (a) show what is known as "minimal creativity," and (b) not be copied directly from someone else's work. *Minimal creativity* means that there must be at least some small portion of the work that is attributable to the author's efforts. A mechanical copying of someone else's idea does not count. For example, if you were to make a cast-iron miniature of the Lincoln Memorial, you would not be able to copyright it, since the statue of Lincoln within it was someone else's work, and long ago fell into the realm of public domain—meaning that *everyone* is free to use its image. Also, the mechanical copying of the statue, showing no creative effort whatsoever, wouldn't meet the originality criterion.

The quality of an author's efforts is not at all an issue in determining the originality of the work. Minimal creativity is just that: *minimal.* All an author has to do is show that his or her work is in some way—in almost *any* way—distinguishable from previously existing works. The new work does not have to be inventive, unique, novel, clever, unusual, or even drastically different from the work of others.

Someone's work is also copyrightable even if it is identical to the work of someone else, provided the author can reasonably show that he or she did not directly copy the work, but rather derived it by means of personal effort.

At this point it seems that the range of what falls under copyright protection is very wide indeed—but let's take a look at some exceptions.

What *Can't* Be Protected?

Although you might have guessed otherwise after reading the preceding information, several categories of material generally are *not* eligible for statutory copyright protection. These include:

- Works that have not been fixed in a tangible form of expression (For example: choreographic works that have not been notated or recorded, and improvisational speeches or performances that have not been written or recorded)
- Titles, names, short phrases, and slogans; familiar symbols or designs; mere variations of typographic ornamentation, lettering, or coloring; mere listings of ingredients or contents

- Ideas, procedures, methods, systems, processes, concepts, principles, discoveries, or devices, as distinguished from a description, explanation, or illustration
- Works consisting *entirely* of information that is common property and contains no original authorship (For example: standard calendars, height and weight charts, tape measures and rulers, and lists or tables taken from public documents or other common sources)
- Works by the U.S. Government

The Public Domain

Other artistic expressions not protected by the copyright laws are those that fall into what is called *the public domain*—in other words, material that is now owned by everyone in general and no one in particular. All of the great literary and artistic works produced over the centuries, and those produced more recently for which the term of the copyright has expired, belong in the public domain. In addition, any material that does not satisfy the three copyright criteria listed earlier also automatically falls into the public domain.

Public domain material may be used by anyone, for any purpose, provided its use is permitted by law. No one—not even the original author, or his or her heirs—may prevent you from using any intellectual property that has fallen into public domain. An interesting aside here: For those of us who work for the government, the 1976 act also denies copyright protection to works produced by federal employees in the course of their duties.

Here's another interesting aside: There are many people who earn a fortune reproducing and selling government information. You've seen them on TV and heard them on the radio, spouting slogans like, "Buy Government Real Estate" or "What to Do to Protect Your Credit." They sell their books for anywhere from twenty to a hundred dollars each, and all they're doing is reproducing noncopyrightable government information.

Works in the public domain on January 1, 1978, remain in the public domain under the current act.

Fact and Fiction

It probably won't surprise you to learn that *facts*, historical *or* contemporary, cannot be copyrighted. That's pretty much common sense. What is *real* is in the public domain by definition. However, it is not obvious to many that often numerous aspects of fictional works also are not protected by copyright. Basic plots, thematic concepts, and certain scenes in plays or novels that necessarily follow from plot situations; and incidents, characters, and settings that are standard in the treatment of a given theme are often treated by the law as ideas, and hence are not considered protected. This is a gray area of the law, however—so, under *some* circumstances, these *same* aspects of creative works *are* protected.

Who *Can* Claim Copyright?

Anything that you create under copyright is your property. Only the author (or those deriving their rights through the author) can rightfully claim a copyright. If you create something legally, it's yours. There is, of course, one very important exception to this rule: *Works for hire,* and one notable variation thereon, *collaborations.* Let's look at the last first.

Collaborations

If a creative work is created by two or more people, by-and-large they become *joint* owners of the copyright. If there is no agreement to the contrary between or among the joint creators, *each* has the right to use the work *and* to allow *others* to use it. Joint creators are not even under any obligation to seek permission for use from the other owners! Each co-owner, however, *is* responsible for seeing that the others receive their share of any profits made. As we'll see in the discussion on licensing, no co-owner can grant an exclusive license without the permission of his or her partners.

Collective Works

Often you'll see an anthology of works by various authors. Magazines are also produced by a variety of contributing artists and writers. Since in such circumstances each piece of work is in some sense separate from the others, it is not considered part of a collaborative effort, but is instead defined as a portion of a *collective* work. In collective works, the copyright to each separate contribution to the work is distinct from the copyright to the collective work as a whole. The publisher or editor might own the copyright to the work *as a whole,* for example, and thus would be able to control its dispensation. The copyright for *each piece,* however, rests initially with the individual author of the contribution.

Works for Hire

If you are the author of a creative work, please read this section very carefully. In the case of a work made for hire, it is the employer, *not the employee,* who is legally considered the author of the work. Work for hire is the only case where the original author has no rights whatsoever, and can't even regain control of his or her work after the thirty-five-year limit has passed. We strongly recommend that you do not *ever* enter into a work-for-hire agreement unless the rewards are exorbitant—because the moment you sign such an agreement, it's all over. The woods are full of authors and artists who wish they hadn't signed a work-for-hire agreement. Even so, we have to admit, work-for-hire is a very common practice.

The work-for-hire statute is in effect only under certain circumstances. Section 101 of the copyright statute defines a "work made for hire" as either of the following:

1. A work prepared by an employee within the scope of his or her employment
2. A work specially ordered or commissioned for use as a contribution to a collective work, as a part of a motion picture or other audiovisual work, as a translation, as a supplementary work, as a compilation, as an instructional text, as a test, as answer material for a test, or as an atlas, if the parties expressly agree in a written instrument signed by them that the work shall be considered a work made for hire

From time to time the work-for-hire statute has been challenged by various artists, although with little success. A great many powerful publishers, motion-picture companies, and other producers would stand to lose a great deal of money without that statute, and so defend it strongly.

To repeat: A work-for-hire situation is the only one wherein someone other than the actual author is unequivocally treated as though he or she *were* the author, for the duration of the copyright.

Minors

Naturally, if someone underage produces a creative work, he or she is fully entitled to claim the copyright on it. However, various state laws may regulate any business dealings involving copyrights owned by minors. For information on the laws pertaining to your own state, the best approach would be to consult an attorney.

Limits on Copyright Protection

While we've already explored a variety of limits on what can and can't be copyrighted, there also are limits on exactly *how much of* a copyrighted work can be protected. Even if you own a legitimate copyright on material, there are still some circumstances that allow others to use it without your permission!

Sections 107 through 119 of the Copyright Act establish limitations on such rights. Some of these limitations specify exemptions from copyright liability—in other words, they limit the conditions under which an author can sue for infringement if his or her work is used without permission. One major limitation is the doctrine of *fair use,* which is given a statutory basis in section 107 of the act. (We will discuss fair use shortly.) In other instances, the limitation takes the form of a *compulsory license,* which will be discussed in our segment on licenses.

Beyond the scope of those two major headings, however, are numerous situations—enumerated in the 1976 act—in which copyrighted work may be used without the permission of, or payment to, the copyright owner. A partial listing of these situations includes:

- The public performance and display of works as part of nonprofit educational teaching activities
- In the course of religious services

- Before a live audience when there is no purpose of direct or indirect commercial advantage
- By airing them on television and radio sets among other types of performance and displays
- The reproduction of pictorial, graphic, and sculptural works in advertising, *under certain circumstances*
- The reproduction and adaptation of works embodied in computer programs, and the reproduction of works by libraries and archives

The circumstances under which each of these limitations is applicable are defined by a set of special rules. They are not complex, but there are many of them that apply to a variety of both different media and different circumstances. While the specifics do not fall within the scope of our general work, if you believe your copyrights have been infringed on, or you wish to use copyrighted material, you should examine these limitations. For a complete listing, consult the highlights of the Copyright Law as reproduced in Copyright Appendix III.

Fair Use

Now we enter into a vague and complex area of copyright law easily understandable in general terms but when down to specifics often a matter of opinion. This is the realm of the fair use doctrine. While its noble intention is to balance the needs of the individual author with the legitimate needs of the public for the free use of information, it is also the limitation most often called on to defend against infringement suits. To make matters even foggier, the term is not directly defined in copyright law.

In place of a direct definition, the statute lists four factors which the courts can apply in determining if the use of copyrighted material has been fair or unfair. The factors are:

1. The purpose and character of the use, including whether such use is of a commercial nature or is for nonprofit educational purposes
2. The nature of the copyrighted work
3. The amount and substantiality of the portion used in relation to the copyrighted work as a whole

4. The effect of the use upon the potential market for or value of the copyrighted work

These four factors have been interpreted by the courts in a variety of ways, depending on the exact circumstances. For example, it is generally understood that reproducing short sections of copyrighted works for the purposes of printing a review of that material is fair use. The key reference in that situation is "short sections." In general, the longer or larger the amount of the copyrighted material reproduced, the less likely the use can be legitimately considered fair. For example, the use of four words and two notes from a lengthy musical work would likely be found to be fair, whereas the printing of a condensed version of a novel without permission would most likely be considered unfair.

The type of work involved is also a major factor. The unauthorized use of works that are primarily information, such as histories, encyclopedias, and documentaries, is less likely to be considered unfair than the use of material intended for entertainment. Condensing a novel without permission, or using a copyrighted melody with new lyrics for an advertising campaign, would most likely not be interpreted as fair use.

Fair use has sometimes been ruled in the following sampling of situations:

1. The educational copying and use of music
2. Parody
3. Incidental use
4. Copies for use of the blind
5. Off-air videotaping for private use

The third (incidental) use refers to the inadvertent or peripheral use of copyrighted material. If a news broadcast is covering an event and a band happens to be playing a copyrighted song in the background, the broadcast of that tape would be considered fair use. Likewise, if a copyrighted statue or painting appeared in the background of a television show or motion picture, and was not in any way the literary or aesthetic subject of the scene, it would most likely be considered fair use—but even *this* is a dicey area.

In general, the courts have been very lenient with parody and satire, but it is generally considered that parodies only have the right to use enough of the copyrighted material to recall the original to mind. Despite the broad interpretation of fair use in the case of parody, there are other dangers to consider. A self-published comic-book satire of Mickey Mouse engendered a court case from The Walt Disney Company that drove the self-publishers into bankruptcy long before the case ever got to court.

A major question in establishing fair use is whether or not the use of the material in any way interferes with the market need for the original. An unauthorized theatrical adaptation of a novel, for example, might diminish the need for either the novel or an authorized stage version of it, or both, and so would be considered unfair use.

We've briefly touched on some of the major parts of this complex issue, but the debate over fair use doubtless is raging in the courts even as you read this. If you are contemplating the use of copyrighted material and you're in any way uncertain as to whether or not your intended use is fair, consult a lawyer, to make sure you're on solid ground before taking another step.

CHAPTER 7

Copyright Notice

With all these exceptions, broad rules, and legal complexities, you may well be asking yourself whether or not copyright protection is worth having at all. Let me assure you that it is, provided you take certain steps to safeguard your protection.

One of the most important steps you can take is to make sure that any printed or otherwise publicly or privately disseminated versions of your work contain a proper copyright notice. Otherwise, someone seeing them may simply assume they are in the public domain, and try to use all or some of them. Aside from informing the public that the work is protected, the copyright notice identifies the copyright owner and shows the year of first publication.

In the event that a work is infringed, if it carries a proper notice the court will not allow a defendant to claim *innocent infringement,* i.e., that he or she did not realize that the work was protected. A successful innocent-infringement claim may result in a reduction in damages that the copyright owner would otherwise receive.

Before March 1, 1989, the use of the copyright notice was mandatory on all published works. Any work first published before that date must bear a copyright notice or risk the loss of copyright protection. For works first published on or after March 1, 1989, the use of the notice is optional but highly recommended.

The use of the copyright notice is the sole responsibility of the *copyright owner*. It does *not* require advance permission from, or registration with, the Copyright Office.

Different Forms for Different Media

Since the material that can be copyrighted can be presented in a variety of ways, the manner of presenting copyright notices varies, depending on the medium or media by means of which the work is expressed. Books, movies, and motion pictures, for example (since they all contain a visual component), fall under the category of *visually perceptible copies*.

Certain other kinds of works—musical, dramatic, and literary, for example—may not be fixed in copies, but by means of sound in an audio recording. Since such audio recordings as tapes and compact discs are phonorecords and are not permitted to be copies, a distinctive symbol is used to indicate protection of the musical, dramatic, or literary work that is recorded: a "P" in a circle: ℗.

Form of Notice for Visually Perceptible Copies

The copyright notice for visually perceptible copies should contain each of the following three elements:

1. The copyright symbol (the letter "C" in a circle: ©), *or* the word "Copyright," *or* the abbreviation "Copr."
2. The year of first publication of the work (In the case of compilations or derivative works that incorporate previously published material, the year date of first publication of the compilation or derivative work is sufficient. The year may be omitted where a pictorial, graphic, or sculptural work, with accompanying textual matter—if any—is reproduced in or on greeting cards, postcards, stationery, jewelry, dolls, toys, or any other useful article.)
3. The name of the owner of copyright, *or* an abbreviation by which the name can be recognized, *or* a generally known alternative designation of the owner

Form of Notice for Phonorecords of Sound Recordings

Sound recordings are defined as "works that result from the

fixation of a series of musical, spoken, or other sounds, but not including the sounds accompanying a motion picture or other audiovisual work, regardless of the nature of the material objects, such as disks, tapes, or other phonorecords in which they are embodied."

The copyright notice for phonorecords of sound recordings has somewhat different requirements. The notice appearing on phonorecords should contain the following three elements:

1. The symbol ℗
2. The year of first publication of the sound recording
3. The name of the owner of the copyright, *or* an abbreviation by which the name can be recognized, *or* a generally known alternative designation of the owner

If the producer of the sound recording is named on the phonorecord labels or containers, and no other name appears in conjunction with the notice, the producer's name shall be considered a part of the notice.

Since questions may arise from the use of variant forms of the notice, any form of the notice other than those given here should not be used without first seeking legal advice.

Position of Notice

The notice should be affixed to copies or phonorecords of the work in such a manner and location as "[to] give reasonable notice of the claim of copyright." In the case of phonorecords, the notice may appear on the surface of the phonorecord, *or* on its label *or* container, provided the manner of placement and location give reasonable notice of the claim. The three elements of the notice should ordinarily appear *together* on the copies or phonorecords.

The Copyright Office has issued specific regulations concerning the form and position of the notice, in the Code of Federal Regulations (37 CFR, Part 201). For more information, request Circular 3.

Works by the U.S. Government

Works by the U.S. Government are *not* eligible for copyright protection. Works that utilize government works in whole or in part, yet claim copyright for the nongovernmental portion of the work, should use a notice that identifies *either* those portions of the work in which copyright is claimed, *or* those that constitute U.S. Government material, *or* both.

An example of both might be:

© 1991 Jane Brown
Copyright claimed in Chapters 7–10,
exclusive of U.S. Government maps

Works published before March 1, 1989, that consist primarily of one or more works of the U.S. Government must bear a notice *and* the identifying statement.

Unpublished Works

If your work is as yet unpublished, in order to prevent someone from inadvertently publishing your work without the proper copyright notice you may wish to place such a notice on any copies or phonorecords that leave your control. The rules for unpublished copyright notices are the same as those that apply for published works.

If the Notice Is Omitted or Incorrect, 1978–1989

Since the 1976 act made such sweeping changes in the copyright law, a grace period for corrections was provided. Sections 405 and 406 of the Copyright Act list procedures for making corrections if the copyright notice either was omitted entirely, or in some manner was incorrect; and the work was published on or after January 1, 1978, but before March 1, 1989.

In general, for works incorrectly copyrighted during the almost decade-long grace period, the copyright is *not* automatically lost. Protection may be maintained if *either* of the following applies:

- The work has been registered with the Copyright Office
- The work is registered within five years after publication with the faulty notice, and a reasonable effort is made to add or correct the notice on all copies or phonorecords

distributed to the public in the United States after the mistake has been discovered

The Copyright Office lists the entire correction procedure in Circular 3.

CHAPTER **8**

Copyright Registration

As we've already discussed, the copyright itself is secured *automatically* when the work is fixed in a copy or phonorecord for the first time. *No* publication, registration, or other action is required to secure the copyright. There are, however, certain definite advantages you acquire when you *do* register your work with the Copyright Office. Among these are the following:

- Registration establishes a public record of the copyright claim.
- Registration is necessary for works of U.S. origin, and for foreign works not originating in a Bern Union country, before an infringement suit may be filed in court.
- If made before or within five years of publication, registration will establish *prima facie* evidence in court of the validity of the copyright.
- If registration is made three months after publication of the work, or prior to an infringement of the work, statutory damages and attorney's fees will be available to the copyright owner in court actions. Otherwise, only an award of actual damages and profits is available to the copyright owner.

When Works Can Be Registered

Registration may be made at any time within the life of the

copyright. Prior to 1978, when a work was registered in unpublished form, it was necessary to make another registration when the work became published. This is no longer the case.

Renewals

Any works published with notice of copyright before January 1, 1978, must be registered and renewed during the first twenty-eight-year term of copyright, in order to maintain protection.

Registration Procedures

Unlike the lengthy and complex patent procedure, which requires the assistance of a patent attorney or agent, the copyright registration process is fairly simple and straightforward. It can be done by anyone, in a short amount of time.

To register your work, assemble the following in the same envelope or package:

1. A properly completed application form
2. A nonrefundable filing fee of $20 for each application
3. A nonreturnable "deposit" (copy) of the work being registered

Send it to:

The Register of Copyrights
Copyright Office, Library of Congress
Washington, DC 20559

The application should be completed using a black-ink pen or typewriter. You may submit photocopies of the application forms, provided they are clear, legible, on a good grade of white paper, and printed head-to-head (i.e., so that when you turn the sheet over, the top of page 2 is directly behind the top of page 1).

In response, you'll receive a certificate of registration which has been partly reproduced directly from your application forms—so it is vital that your forms meet these requirements. Forms *not* doing so will be returned.

Many people confuse the copyright certificate registration number with a Library of Congress Catalog Number. They are not the same thing. For more information on the Library of Congress Catalog number and how to obtain one, please refer to Copyright Appendix IV.

Deposit Requirements

As we have just seen, *deposit* (in this sense) doesn't refer to money, but rather to the copy—or copies—of the work to be registered for copyright. The requirement for the deposit—that is, the number of copies—varies from situation to situation, as follows:

- If the work is unpublished, *one* complete copy or phonorecord
- If the work was first published in the United States on or after January 1, 1978, *two* complete copies or phonorecords of the best edition
- If the work was first published in the United States before January 1, 1978, *two* complete copies or phonorecords of the work as first published
- If the work was first published outside the United States, whenever published, *one* complete copy or phonorecord of the work as first published

Special Deposit Requirements

Certain types of works call for special deposit requirements. Sometimes a complete copy isn't necessary at all, and only identifying material is required. In still other instances, the deposit requirement may be unique. The following are three prominent examples of exceptions to the general deposit requirements.

- If the work is a motion picture, the deposit requirement is one complete copy of the unpublished *or* published motion picture *and* a separate written description of its contents—such as a continuity, press book, or synopsis.
- If the work is a literary, dramatic, or musical one pub-

lished only on phonorecord, the deposit requirement is one complete copy of the phonorecord.

- If the work is an unpublished or published computer program, the deposit requirement is one visually percep-tible copy in source code of the first and last twenty-five pages of the program. For a program of fewer than fifty pages, the deposit is a copy of the entire program. (For more information on computer program registration, including deposits for revised programs and special relief for trade secrets, request Circular 61.)

In the case of works reproduced in three-dimensional copies, such as sculptures, only identifying material such as photographs or drawings is ordinarily required. Other exam-ples of special deposit requirements include

- Greeting cards
- Toys
- Fabric
- Oversized material
- Video games
- Automated data bases
- Contributions to collective works

This list is by no means exhaustive, so if you are unsure of the deposit requirement for *your* work, write or call the Copyright Office and describe the work you wish to register.

Unpublished Collections

A work may be registered in unpublished form as a *collection*, with one application and one fee, under the following conditions:

- The elements of the collection are assembled in an orderly form.
- The combined elements bear a single title identifying the collection as a whole.
- The copyright claimant in all three elements and in the collection as a whole are the same.
- All of the elements are by the same author—or, if they are by different authors, at least one of the authors has contributed copyrightable authorship to each element.

While all registered works are indexed in the Copyright Office's *Catalog of Copyright Entries,* a collection is indexed only under the collection title—not under the titles of the individual works that the collection comprises—unless they have been otherwise registered.

Filing Corrections

To correct a copyright registration or add to the information in a registration, you would have to file a supplementary registration form, Form CA. A variety of forms from the Copyright Office serve different functions. A complete list of these forms and their purposes appears in Copyright Appendix I.

The information in a Form CA may *augment,* or add to (but not supersede, or substantially alter) the information contained in the earlier registration. A supplementary registration is *not* a substitute for either an original registration or a renewal registration, *or* for recording a transfer of ownership. For further information about supplementary registration, request Circular 8.

Mandatory Deposit for Works Published in the United States

Although copyright registration is not required, the Copyright Act establishes a mandatory deposit requirement for works published in the United States. In general, the owner of the copyright, or the owner of the exclusive right of publication in the work, has a legal obligation to deposit, within three months of publication in the United States, two copies of the work (or, in the case of sound recordings, two phonorecords) for the use of the Library of Congress. Failure to make the deposit can result in fines and other penalties. It is important to note, however, that the fines or penalties do *not* affect copyright protection.

Certain categories of works are exempt from the mandatory deposit requirements; for others, the obligation is reduced. For further information about mandatory deposit, request Circular 7d.

Who May File an Application Form

There is *no* requirement that applications be prepared or filed by an attorney. In fact, *any* of the following persons is legally entitled to submit an application form:

The Author: This is either the person who actually created the work, or, if the work was made for hire, the employer or other person for whom the work was prepared.

The Copyright Claimant: This party is defined in Copyright Office regulations as either the author of the work, or a person or organization that has obtained ownership of all the rights under the copyright initially belonging to the author. This category includes a person who or organization which has obtained by contract the right to claim legal title to the copyright in an application for copyright registration.

The Owner of Exclusive Right(s): Under the law, any of the exclusive rights that go to make up a copyright, and any subdivision of them, can be transferred and owned separately, even though the transfer may be limited in time or place of effect. The term "copyright owner," with respect to any one of the exclusive rights contained in a copyright, refers to the owner of that particular right. Any owner of an exclusive right may apply for registration of a claim in the work.

The Duly Authorized Agent of Such Author, Other Copyright Claimant, or Owner of Exclusive Right(s): Any person authorized to act on behalf of the author, other copyright claimant, or owner of exclusive rights may apply for registration.

Mailing Instructions

All applications and materials related to copyright registration should be addressed to the Register of Copyrights, Copyright Office, Library of Congress, Washington, DC 20559. The application, nonreturnable deposit (copies, phonorecords, or identifying material), and nonrefundable filing fee should be mailed in the same package.

What Happens If the Three Elements Are Not Received Together

Applications and fees received without appropriate copies, phonorecords, or identifying material will *not* be processed, but rather will ordinarily be *returned*. Unpublished deposits without applications or fees will also ordinarily be returned. In most cases, published deposits received without applications and fees can be immediately transferred to the collections of the Library of Congress. This practice is in accordance with Section 408 of the law, which provides that the published deposit required for the collections of the Library of Congress may be used for registration "[only if the deposit is] accompanied by the prescribed application and fee."

After the deposit is received and transferred to another department of the Library for its collections or other disposition, it is no longer available to the Copyright Office. If you wish to register the work, you must deposit additional copies or phonorecords with your application and fee.

Fees

Do not send cash! A fee sent to the Copyright Office should be in the form of a money order, check, or bank draft payable to the Register of Copyrights. It should be securely attached to the application. A remittance from outside the United States should be payable in U.S. dollars and should be in the form of an international money order, or a draft drawn on a U.S. bank. Do not send a check drawn on a foreign bank!

Effective Date of Registration

A copyright registration is effective on the date the Copyright Office receives all of the required elements in acceptable form, regardless of how long it takes to process the application and mail the certificate of registration. The time the Copyright Office takes with an application varies, depending on the amount of material the Office is receiving and the personnel available to handle it. Keep in mind that it may take a number of days both for mailed material to reach the Copyright Office, and for the certificate of registration to reach the recipient after being mailed by the Copyright Office.

What to Expect From the Copyright Office

If you are filing an application for copyright registration in the Copyright Office, you will *not* receive an acknowledgement that your application has been received. What you *can* expect is:

- A letter or telephone call from a copyright examiner *if* further information is needed
- A certificate of registration to indicate the work *has* been registered
- A letter explaining why registration has been refused, if it *cannot* be made

Allow about 120 days to receive a letter or certificate of registration. If you want to know when the copyright Office receives your material, send it by registered or certified mail and request a return receipt from the post office. Allow at least three weeks for the return of your receipt.

Search of Copyright Office Records

The records of the Copyright Office are open to the public for inspection and searching. You may ask yourself, "Why would I want to search through copyright records?" Well, there are *several* reasons. For one thing, you might want to determine whether someone owns an earlier copyright on work similar to your own. (If the similarity is great enough, that might make your own copyright invalid!) Or perhaps you suspect that your copyright has been infringed by someone else's work. And yet another reason is simple inspiration: A trip through the masses of artistic works in the copyright files might be just the thing to get your own creative juices flowing.

On request, the Copyright Office will search its records, for a charge of $20 per hour or fraction thereof. For information on searching the Office records concerning the copyright status or ownership of a work, request Circulars 22 and 23.

How Long Copyright Protection Endures

As we have at least strongly hinted, this is another of the instances whereby the 1976 Act created some major alterations. The rules that apply for copyright duration are different for works copyrighted on or after January 1, 1978, and those copyrighted before. Let's see just how different.

Works Originally Copyrighted on or After January 1, 1978

A work created on or after January 1, 1978, is automatically protected from the moment of its creation. The term of its copyright protection ordinarily lasts for the duration of the author's life, plus an additional fifty years. In the case of a *joint work*—one prepared by two or more authors who did not do work-for-hire—the term lasts for fifty years after the last surviving author's death. In the case of works made for hire, or for anonymous and pseudonymous works (unless the author's identity is revealed in Copyright Office records), the duration of the copyright is seventy-five years from publication or one-hundred years from creation, whichever is shorter.

Works that were created but not published or registered for copyright before January 1, 1978, have been automatically brought under the statute and are now given federal copyright protection. The duration of copyright in these works is generally computed in the same way as for works created on or after January 1, 1978.

The law provides that in *no* case will the term of copyright for works in this category expire before December 31, 2002; and that for works published on or before December 31, 2002, the term of copyright will not expire before December 31, 2027.

Works Copyrighted Before January 1, 1978

Under the law in effect before 1978, the copyright was secured either on the date a work was published or on the date of registration, whichever came first. In either case, the copyright endured for a *first* term of twenty-eight years from the date it was secured. During the last (twenty-eighth) year of the first term, the copyright was eligible for renewal. *Current* copyright law extends the renewal term from twenty-eight to forty-seven years for copyrights existing on January 1, 1978, making these works eligible for a total term of protection of seventy-five years.

However, the copyright must be *renewed* to receive the forty-seven-year period of added protection. This is accomplished by filing a properly completed Form RE, accompanied by a $12 filing fee, with the Copyright Office before the end of the twenty-eighth calendar year of the *original* term.

For more detailed information on copyright terms, write to the Copyright Office and request Circulars 15a and 15t. For information on how to search the Copyright Office records concerning the copyright status of a work, request Circular 22.

CHAPTER 9

Publication and Licensing

Although it no longer is the key to obtaining copyright protection, publication remains important to copyright owners—for several reasons by-and-large already discussed. Let's summarize them:

- When the work is published, the copyright notice identifies the year of publication and the name of the copyright owner. This informs the public that the work is protected by copyright. Works published before March 1, 1989, must bear the notice or risk loss of copyright protection.
- Works published in the United States are subject to mandatory deposit with the Library of Congress.
- Publication of a work can affect the limitations on the exclusive rights of the copyright owner.
- The year of publication may determine the duration of protection for both anonymous and pseudonymous works (i.e., works the identity of whose author or authors is not revealed in the records of the Copyright Office) and for works made for hire.
- Deposit requirements for registration of published works differ from those for registration of unpublished works.

What Is Publication?

It may seem all too obvious to many of us precisely what constitutes publication, but in legal terms the definition has a much more complex and specific meaning. The Copyright Act itself defines publication as:

> [The] distribution of copies or phonorecords of a work to the public by sale or other transfer of ownership, or by rental, lease, or lending. The offering to distribute copies or phonorecords to a group of persons for purposes of further distribution, public performance, or public display, constitutes publication. A public performance or display of a work does not of itself constitute publication.

More about the definition of "publication" can be found in the legislative history of the Act. Legislative reports define "to the public" as a "distribution to persons under no explicit or implicit restrictions with respect to disclosure of the contents."

The reports further state that this definition clarifies the fact that the sale of phonorecords constitutes publication of the underlying work (i.e., the musical, dramatic, or literary work embodied in a phonorecord). Even when copies or phonorecords are offered for sale or lease to a group of wholesalers, broadcasters, or motion-picture theaters, publication *does* take place—but only if the purpose is *further* distribution, public performance, or public display. On the other hand, any form of dissemination in which the material object *does not* change hands—for example, performances or displays on television—*doesn't* constitute publication, no matter how many people are exposed to the work.

How to Proceed With Your Copyrighted Work

Up until now, we've been talking about the type of protection offered by the copyright, and how to go about securing it. Since the method of obtaining copyright registration is so simple, there has been no need (as yet) to discuss hiring lawyers, agents, or any third party. Whether you've written a treatise on whale hunting or sculpted a giant chocolate figure, you fill out the same forms and receive the same protection.

Now, of course, the question becomes, "What next?" Because copyright, unlike patents, covers such a wide area of material, it is almost impossible to address each one. In the realm of inventing, there is a basic similarity in the representation and marketing of products. Patent agents keep in touch with a variety of industries, from toy companies to stationery supply houses. Not so with copyright property!

Let's look at just three types of work that fall under copyright protection: music, literature, and computer programs. In each case the handling, marketing, and selling involve radically different markets, industries, methods of distribution, and even retail outlets. While we will be able to provide some general suggestions for those areas that cover all creative works, we strongly suggest you do a great deal of research involving the particular industry you wish to work in.

Is an Agent Necessary?

A good agent, knowledgeable in the field and devoted to your work, is *invaluable*. However, unlike as with inventions, with creative work it's a bit easier to have some success on your own. There's a very simple reason for this—and it's *not* because people are more open to art. Look at it this way: If you sell a new invention to a going business, they have to hire employees, find the supplies to manufacture it, retool machinery, create different molds, and so on. All of this often costs millions of dollars.

Now let's say you own a small record company, publishing venture, or art gallery. You already own your sound equipment, your desktop publishing stuff, or your gallery. If *you* acquire a new "invention," it's within a realm *not* requiring retooling and such. Also, your mailing list or distribution system already is in place. And so on. Obviously there still will be expenses, but those will stay in the tens of thousands max, not the millions. To put it simply, in the artistic realms there often is less capital at risk—so people are at least *slightly* more willing to risk theirs. (Maybe even a *lot* more.) On the down side, this means there are many more small companies than big ones going in and out of business every year. So always be careful, when you make a deal, that the business you're getting involved with is on solid financial footing.

If you've got a new novel (or whatever), it might well be worth your while to visit a book store, find out who's publishing work similar to your own, and then present those publishers with your work. (Submission procedures vary from publisher to publisher, so write or call first and ask for particulars.) Better yet, be aware that publishers' names, addresses, phone numbers, and product markets are listed in annually updated reference books you can ask any librarian about, and often can buy through bookstores. The one most used in the publishing business is Bowker's *Literary Market Place*. Many major publishers will refuse to read unsolicited manuscripts not presented by an agent. If the publishers you've singled out as your best shots fall into that category, you should contact them to see if you can find out which literary agents they work with regularly, then approach at least one of those agents for possible representation. Unsolicited and unagented manuscripts often are collected in what's known as a "slush pile." Occasionally—*very* occasionally—a publisher will actually give the slush pile a look. But don't bet on the fate of your *own* opus if you send it in "blind."

Getting an Agent Anyway

There are three valid reasons for seeking representation for your work. The first is that, rather than collecting royalty payments, you might be building up a pile of rejection letters. Aside from your going out and taking classes in your field, you can get from an agent helpful critiques, not only about what's wrong with your work but about what's really hot in the market. The second reason is that an agent with contacts can get your material seen by the people who *count*. The third and final reason is that once you sell your work, an agent with knowledge in the field can often negotiate a better contract for your next book. A good agent will know when and how to push for more, and when to sign on the dotted line. Too, an agent can often be hired by an author even *after* a company becomes interested in an author's work—purely to act as a negotiator! In such cases, the agent will sometimes charge a lower commission.

Good agents, like good patent agents, attorneys, doctors, and friends, are hard to find—but you might start with the phone book, or contact fellow authors for recommendations. Many agents charge a minimal fee to review work. Beyond these minimal fees, beware of any agent who asks for money up front. What you want is someone who makes a living from selling your *work,* not from selling you on a *scam.* Also keep in mind that just because you have a copyright on something doesn't mean you'll be able to interest an agent in your work. Since agents work on a percentage, they *have* to believe they can sell what you have to offer. In addition, many legitimate agents (in fact, almost all of the larger agents) charge a small up-front "reading fee" that can be as high as several hundred dollars, before they'll even look at your work.

Producing an Attractive Package for Your Work

The old axiom about artists is that they generally suffer in poverty during their lifetime, only to become famous after they die. This may be because they never mastered the art of packaging.

At some point or another, you're going to have to try to convince someone to look at your work and consider investing in it. This is generally done through the mail, with a query letter for books, or a demo tape for musicians, or a portfolio for photographers. Whereas content is the ultimate deciding factor, how your samples are presented—whether to an agent, a producer, or a publisher—is a key to your success. It's sad to consider, but it's possible that if someone who is lousy creatively can put together a great presentation package, he or she may get more attention and more money than a brilliant artist with a lousy presentation package.

Anything that will get your package pulled out from the pile and noticed is good. A captivating cover letter, even better-written than the novel it represents, or a poster-size blow-up of one of your paintings, are just two examples. The package should also be succinct and attractive.

One of the most important things to remember when preparing your package is that you're not trying to summarize your work, or even necessarily capture its essence. You're trying to get someone who doesn't have much time to *take a good look* at what you've got to offer. You are, in effect, writing a commercial or an advertisement for yourself. You want the reader glued to your presentation for the short time it should last, and you want him or her to end wanting to see more. A good idea might be to take a look at an advertising campaign you found particularly appealing, and think about what made it work—then incorporate some of those principles into your presentation.

But what's even more important is *networking*. The more people you get to know, the better. You might even have reinvented the wheel—but if you don't know whom to roll it to, eventually it's going to go flat.

Types of Agreements

As we've said before, if you sign a work-for-hire agreement and produce work under it, the negotiations are, in effect, completed. Your employer owns the copyright, and can generally do whatever he or she wishes with the work—even without consulting you. If you've retained your copyright, however, there are two major options in making deals. The first is licensing your work, which transfers certain rights in your work to a company for a certain period of time, in exchange for remuneration. The second is a total sale of your copyright, giving a company the right to be legally considered as the author. As was the case with patents, I do not recommend selling your copyright outright, unless you are offered a phenomenal fee. Licensing, or its equivalent, allows for greater long-term income and, more important, allows you to retain at least *some* control over your work.

Transfer of Copyright

If you have an offer from a company to publish or otherwise reproduce your work, you'll be required to transfer certain rights to them. For example, in order for a magazine to publish an article by you, they must first acquire that right from you, in exchange for either a flat fee or a royalty—the latter a percentage of the money they make from the sale of your work. Much in keeping with the spirit of the law, mere *ownership* of a book, manuscript, painting, or any other copy or phonorecord does *not* give the possessor the copyright. And the transfer of ownership of a material object that embodies a protected work does *not* of itself convey any rights in the copyright.

What Can Be Transferred

Any or all of the exclusive rights, or any subdivision of the rights, of the copyright owner may be transferred, but the transfer of exclusive rights is *not* valid unless that transfer is in writing and signed by the owner of the rights conveyed (or such owner's duly authorized agent). The transfer of a right on a nonexclusive basis does *not* require a written agreement.

A copyright may also be conveyed by operation of law, and may either be bequeathed by will, or pass as personal property by the applicable laws of intestate succession. Copyright is a *personal* property right, and as such is subject to the various state laws and regulations that govern the ownership, inheritance, or transfer of personal property, as well as terms of contracts and conduct of business. For information about relevant state laws, consult an attorney.

Types of Transfer Contracts

As we've seen, the owner of a copyright may not be in the best position to exploit his or own work—and, as a result, may enter into an agreement with a person or business that for their mutual financial benefit *can*. An agreement from one person that gives another permission to use all or part of a copyright is called a *copyright license*. Copyright licenses fall into two general groups, *exclusive* and *nonexclusive*. In the former case, the licensee becomes the *only* entity entitled to exploit the author's work; in the latter, the author is free to sign numerous deals with numerous entities, even if those entities would then be in direct competition with each other.

Exclusive Licenses

The owner of an exclusive license exerts a monopoly over the creative work for the designated term of the license. It is, in effect, treated as a "transfer" of copyright ownership, entitling the license owner to all the rights and privileges of the author, essentially giving him or her a free hand in utilizing the rights.

An *exclusive* licensee is entitled to:

- Register for copyright in the work
- Institute infringement proceedings when applicable
- List himself or herself as owner in the copyright notice

Nonexclusive Licenses

Each of the rights in the work may be separately licensed by the owner, independent of the other rights. Each right can also be subdivided, and each subdivision licensed. To an extent, the imagination of the parties involved is the only limiting factor regarding how and when rights may be exercised on the license.

Naturally, the rights of the nonexclusive-license owner are restricted by the agreement made with the author. A *nonexclusive* licensee generally:

- Cannot list himself or herself in the copyright notice as the author
- Cannot institute infringement proceedings

Contract Points

All licensing agreements should be in writing, and any transfer agreements should be registered with the Copyright Officer. In addition, contracts should address the following points:

- The term of the agreement
- Whether the license is exclusive or nonexclusive
- A warranty by the author (or "licensor") that the work is original
- Whether the work has been registered or not
- A description of the rights covered by the agreement
- An obligation of the "licensee" to maintain accurate books on royalty payments, to make payments at specific times, and to include a written statement
- The right of the licensor to inspect the licensee's books
- An agreement from the licensee to indemnify the licensor from any lawsuits resulting from the licensee's use of the rights
- An agreement from the licensor to indemnify the licensee from any lawsuits resulting from a breach of the licensor's warranty of originality

- An agreement on the part of the licensee to carry insurance to protect the licensor from any claims asserted against the licensor as a result of the licensee's acts
- An agreement on the part of the licensor to cooperate with the licensee in the event the rights are infringed
- An agreement on the part of the licensor to protect and defend the licensed rights
- Provisions for termination in the event of default on the part of either party
- Provisions for the serving of notices and payment of royalties
- A prohibition on the right of one or both parties to reassign to a third party any rights or obligations
- An indication of the state or jurisdiction where a lawsuit involving controversies may be filed
- A representation that those signing the agreement have the authority to do so
- A statement that the licensee will execute all documents the licensor may reasonably require in order to effect a termination of the license
- A statement that the contract constitutes the entire agreement between the parties
- A statement that the agreement is binding not only on the parties but also on their parent companies, subsidiaries, related companies, affiliates, legal representatives, heirs, assigns, and successors
- An obligation on the part of the licensee to use all copyright notices designated by the licensor on all copies of the work produced, and an acknowledgment from the licensee that compliance with this obligation is a condition of the licensee having the authority to exercise the rights covered by the license
- A guarantee to produce some sort of income from the license within a certain amount of time, or a return of the rights to the author

For more information, refer to the sample license contract that appears in Copyright Appendix II of this section.

Compulsory Licensing

Most often, licenses are negotiated, and the copyright owner is free to withhold permission to use his or her work. As we discussed earlier, under the fair use statutes it is possible, given certain circumstances, to use copyrighted work without either the permission of the author or remuneration. It may surprise you to learn, however, that under certain other circumstances, the copyright owner is *compelled by law* to give permission. This is referred to as *compulsory licensing*. In the case of a compulsory license, the author is still entitled to a preset royalty fee for the use of his or her work; the only difference is that permission cannot be denied. Only *published works* are subject to compulsory licensing.

There are two major categories of works which can be compulsorily licensed: musical works that have already been released to the public, and any works publicly performed.

In the first instance, if someone wishes to rerecord a work that has been "published," he or she may do so *without* the permission of the author, provided the rerecorder pays the appropriate fees. In the second instance—public performances—a television or radio station may likewise broadcast a live performance of a work *without* first requesting permission; again, subject to set royalty payments. This is referred to as a *secondary transmission*.

There are four specific types of compulsory licenses set forth in the 1976 act:

1. Making and distributing phonorecords embodying musical work
2. Publicly performing music on jukeboxes
3. Making secondary transmissions by cable systems
4. Publicly performing works by noncommercial broadcasters

The terms and conditions of compulsory licenses are not negotiable. In some cases, the license fees payable under them are collected by the Copyright Office or one of several professional organizations, such as ASCAP or the Writers or Directors Guild. They are then distributed to the owners of the rights.

Termination of Transfers — The Thirty-five-Year Limit

Under previous law, regardless of any transfer agreements, exclusive *or* nonexclusive, the copyright in a work reverted to the author, if living (or if the author was *not* living, to other specified beneficiaries), provided a renewal claim was registered in the twenty-eighth year of the original term. Under present law, the renewal is no longer necessary—except for works already in the first term of protection when the present law took effect.

According to the present law, after thirty-five years (under most circumstances), *any* transfer of copyright may be made invalid, and the author may reclaim full control of his or her copyright by serving written notice on the transferee—within specified time limits. The biggest exception to this rule is work-for-hire, in which the original author has *no* copyright claims.

For works under copyright protection before 1978, the present law provides a similar right of termination, covering the extended years of the former maximum term from fifty-six to seventy-five years. For further information, request Circulars 15a and 15t.

Recording Transfer Contracts

Transfers of copyright are normally made by contract. The Copyright Office does not have or supply any forms for such transfers. However, transfers of copyright ownership can be recorded in the Copyright Office. Although such recording is not required to make the transfer valid, it does provide certain legal advantages, and may be required to validate the transfer against third parties. For information on the recording of transfers and other documents related to copyright, request Circular 12.

Infringement

In the simplest of terms, *infringement* is an unauthorized copying of your work that falls outside the doctrine of fair use and also of compulsory licensing. Infringement can run the gamut from a publisher's issuing of an unauthorized edition of your protected work, to someone paraphrasing your material in a work of his or her own. The first instance is a clear-cut example of infringement. The second, however, is open to subtle and complex interpretation by the courts.

If it ever comes to pass that you believe your protected work is being unfairly used—or infringed upon—by another, there are several steps you should take. The first is to determine whether or not infringement actually *is* taking place. You can get a fast read on that by reviewing our earlier section on fair use and making certain that the utilization of your work doesn't fall into any of the categories outlined there.

Next (if you haven't done so already), ensure that your work has been *registered*. It is here that the registration certificate, validating the date of registration and creation of your work, comes in most handy. If you're dealing with a direct, verbatim copy of a substantial portion of your work, your next step would be to send the infringer a "Cease and Desist" notice outlining the specifics of the infringement, requesting that all use of your work stop, and insisting on fair remuneration. In addition to putting the infringer on notice, this also creates a "paper trail" of your efforts to protect your rights; such evidence would prove useful if the case goes to court. You should also know that once you become aware of an infringement, you have a three-year limit during which you can bring charges.

If you're dealing with a paraphrase of your work, you should proceed more cautiously. Make a list of similarities between the works. In the case of a written work, this doesn't necessarily have to consist only of verbatim copying; flavor, story details, and nuances are also important. None of these items alone will prove out an infringement suit—but the longer your list, the more substantial your claims become.

Next, see if you can show that the infringer somehow gained access to your work. It isn't necessary to *prove* access, but creating the suspicion of it does tend to build a strong case. Once you have your information clearly sorted out, compose a strong cease-and-desist letter, preferably with the help of a lawyer. If the letter fails to work, your next step would be (if you could afford it) to retain a lawyer and take your case to court. A successful infringement suit could bring you appropriate remuneration for the illicit use of your work, and an injunction against the infringer (to protect your work from *further* infringement).

Keep in mind that the shoe could also appear suddenly on the other foot: Someone could accuse *you* of infringing on *his or her* copyright. To safeguard against such instances, make certain that your own use of other people's work clearly falls under the fair use doctrine. Beyond that, it's *usually* a good idea to seek the permission of the author of the work you are using. Even if you think it's in the public domain or that it's fair use, you can avoid a lot of potential problems by getting permission in writing. Also, listen to your own instincts. If you feel as though you're stealing someone else's work, chances are you probably are.

The Future of Copyright Law

As we saw in our opening sections, the copyright laws of the United States have been in a perpetual state of evolution. New modes of creation and new methods of distributing those creations have elicited changes and variations in the law. Now, in the computer age, the copyright laws are facing perhaps their greatest challenge: digitized information.

The fast-paced advances in personal computer technology are rapidly outpacing the old intellectual property laws. With the advent of the CD–ROM drive, entire libraries can be transmitted throughout the world via communications media and printed out by any properly equipped compatible unit. In the era of audio and video recording, digital equipment creates copies that are at least as good as the original. The problem then becomes: If an individual can store, send, and receive massive amounts of supposedly "protected" material quickly and easily, how could one possibly keep track of the licensing fees?

To date, this conundrum has placed some major industries in the unenviable position of trying to hold back new technology, at least until they can figure out a way to protect their work. Obviously, with high-tech industry growing throughout the world, this won't work for long. Major publishers and record and movie producers are currently working with computer companies to arrive at some general licensing fee or "level of copying" that everyone can live with. Changes in technology were what in large part brought about the 1976 act. Watch for further sweeping revisions of the copyright laws in the near future.

A List of Copyright Forms

For Original Registration

Form TX: For published and unpublished nondramatic literary works.

Form SE: For serials—works issued, or intended to be issued, in successive parts bearing numerical or chronological designations and intended to be continued indefinitely (periodicals, newspapers, magazines, newsletters, annuals, journals, etc.).

Short Form/SE and Form SE/GROUP: Specialized SE forms for use when certain requirements are met.

Form PA: For published and unpublished works of the performing arts (musical and dramatic works, pantomimes and choreographic works, motion pictures, and other audiovisual works).

Form VA: For published and unpublished works of the visual arts (pictorial, graphic, and sculptural works, including architectural works).

Form SR: For published and unpublished sound recordings.

For Renewal Registration

Form RE: For claims to renewal copyright in works copyrighted under the law in effect through December 31, 1977 (1909 Copyright Act).

For Corrections and Amplifications

Form CA: For supplementary registration to correct or amplify information given in the Copyright Office record of an earlier registration.

For a Group of Contributions to Periodicals

Form GR/CP: An adjunct application to be used for registration of a group of contributions to periodicals in addition to an application Form TX, PA, or VA.

Free application forms are supplied by the Copyright Office.

Copyright Office Hotline

Requestors may order application forms and circulars at any time by telephoning (202) 707-9100. Orders will be recorded automatically and filled as quickly as possible.

Copyright Appendix II
Copyright Forms

The Copyright Fees and Technical Amendments Act of 1990 (Public Law 101-318) amends the Copyright Act of 1976 by increasing fees for Copyright Office services, effective January 3, 1991. Citations below are to sections of the Copyright Act of 1976, as amended by Public Law 101-318.

	Fees effective 1/3/91	
REGISTRATION OF COPYRIGHT CLAIMS (Forms TX, VA, PA, SR, CA, SE or GR/CP) Form SE/Group (minimum fee $20) Form MW Form RE	$20.00 $10/issue $20.00 $12.00	For each registration and renewal you will receive a certificate bearing the Copyright Office seal.
RECORDATION OF DOCUMENTS Recordation, under section 205, of a document of six pages or less listing no more than one title. Additional titles: each group of 10 or fewer	$20.00 $10.00	A document which relates to any disposition of a copyright, such as a transfer, will, or license, may be recorded in the Copyright Office. When processing is completed, the submitted document(s) will be returned to you, along with a certificate of recordation for each document.
CERTIFICATIONS Additional certificates: each Any other certification: each	$8.00 $20/hr. or fraction	Certified copy of the record of registration, including certifications of Copyright Office records. NOTE: fees are cumulative: certification fees are in addition to any other applicable fees, i.e., search, photoduplication, etc.
SEARCHES Reports from official records: per hour or fraction. Locating Office records: per hour or fraction	$20.00 $20.00	The Copyright Office will, upon request, estimate the fee required for a search; the fee must be received before the search is undertaken.
FILING OF NOTICE OF INTENTION TO MAKE AND DISTRIBUTE PHONORECORDS	$12.00	For the filing, under section 115(b), of notice of intention to make and distribute phonorecords.
RECEIPT FOR DEPOSITS each receipt	$4.00	For the issuance under section 407, mandatory deposit for the Library of Congress, of a receipt for deposit.
SPECIAL HANDLING FEE Registration (plus registration or renewal fee) (see above) Additional Fee For each claim given special handling if a single deposit copy covers multiple claims and special handling is requested only for one. This charge may be avoided by submitting a separate deposit copy. Recordation of a Document (plus recordation fee)(see above)	$200.00 $50.00 $200.00	Special handling is granted at the discretion of the Register of Copyrights in a limited number of cases as a service to those who have compelling reasons for the expedited service. For further information on special handling, you may call (202) 707-9100 and record your request for ML-319 (registration of claims) or ML-341 (recordation of documents).
FULL-TERM RETENTION OF COPYRIGHT DEPOSITS For the full-term retention of copyright deposits under section 704(e).	$135.00	For information on full-term retention, you may call (202) 707-9100 and record your request for Circular 96, Section 202.23.

⊘Filling Out Application Form VA

Detach and read these instructions before completing this form.
Make sure all applicable spaces have been filled in before you return this form.

─BASIC INFORMATION─

When to Use This Form: Use Form VA for copyright registration of published or unpublished works of the visual arts. This category consists of "pictorial, graphic, or sculptural works," including two-dimensional and three-dimensional works of fine, graphic, and applied art, photographs, prints and art reproductions, maps, globes, charts, technical drawings, diagrams, and models.

What Does Copyright Protect? Copyright in a work of the visual arts protects those pictorial, graphic, or sculptural elements that, either alone or in combination, represent an "original work of authorship." The statute declares: "In no case does copyright protection for an original work of authorship extend to any idea, procedure, process, system, method of operation, concept, principle, or discovery, regardless of the form in which it is described, explained, illustrated, or embodied in such work."

Works of Artistic Craftsmanship and Designs: "Works of artistic craftsmanship" are registrable on Form VA, but the statute makes clear that protection extends to "their form" and not to "their mechanical or utilitarian aspects." The "design of a useful article" is considered copyrightable "only if, and only to the extent that, such design incorporates pictorial, graphic, or sculptural features that can be identified separately from, and are capable of existing independently of, the utilitarian aspects of the article."

Labels and Advertisements: Works prepared for use in connection with the sale or advertisement of goods and services are registrable if they contain "original work of authorship." Use Form VA if the copyrightable material in the work you are registering is mainly pictorial or graphic; use Form TX if it consists mainly of text. NOTE: Words and short phrases such as names, titles, and slogans cannot be protected by copyright, and the same is true of standard symbols, emblems, and other commonly used graphic designs that are in the public domain. When used commercially, material of that sort can sometimes be protected under state laws of unfair competition or under the Federal trademark laws. For information about trademark registration, write to the Commissioner of Patents and Trademarks, Washington, D.C. 20231.

Architectural Works: Copyright protection extends to the design of buildings created for the use of human beings. Architectural works created on or after December 1, 1990, or that on December 1, 1990, were unconstructed and embodied only in unpublished plans or drawings are eligible. Request Circular 41 for more information.

Deposit to Accompany Application: An application for copyright registration must be accompanied by a deposit consisting of copies representing the entire work for which registration is to be made.

Unpublished Work: Deposit one complete copy.

Published Work: Deposit two complete copies of the best edition.

Work First Published Outside the United States: Deposit one complete copy of the first foreign edition.

Contribution to a Collective Work: Deposit one complete copy of the best edition of the collective work.

The Copyright Notice: For works first published on or after March 1, 1989, the law provides that a copyright notice in a specified form "may be placed on all publicly distributed copies from which the work can be visually perceived." Use of the copyright notice is the responsibility of the copyright owner and does not require advance permission from the Copyright Office. The required form of the notice for copies generally consists of three elements: (1) the symbol "©", or the word "Copyright," or the abbreviation "Copr."; (2) the year of first publication; and (3) the name of the owner of copyright. For example: "© 1991 Jane Cole." The notice is to be affixed to the copies "in such manner and location as to give reasonable notice of the claim of copyright." Works first published prior to March 1, 1989 , must carry the notice or risk loss of copyright protection.

For information about notice requirements for works published before March 1, 1989, or other copyright information, write: Information Section, LM-401, Copyright Office, Library of Congress, Washington, D.C. 20559-6000.

─LINE-BY-LINE INSTRUCTIONS─
Please type or print using black ink.

1 SPACE 1: Title

Title of This Work: Every work submitted for copyright registration must be given a title to identify that particular work. If the copies of the work bear a title (or an identifying phrase that could serve as a title), transcribe that wording *completely* and *exactly* on the application. Indexing of the registration and future identification of the work will depend on the information you give here. For an architectural work that has been constructed, add the date of construction after the title; if unconstructed at this time, add "not yet constructed."

Previous or Alternative Titles: Complete this space if there are any additional titles for the work under which someone searching for the registration might be likely to look, or under which a document pertaining to the work might be recorded.

Publication as a Contribution: If the work being registered is a contribution to a periodical, serial, or collection, give the title of the contribution in the "Title of This Work" space. Then, in the line headed "Publication as a Contribution," give information about the collective work in which the contribution appeared.

Nature of This Work: Briefly describe the general nature or character of the pictorial, graphic, or sculptural work being registered for copyright. Examples: "Oil Painting"; "Charcoal Drawing"; "Etching"; "Sculpture"; "Map"; "Photograph"; "Scale Model"; "Lithographic Print"; "Jewelry Design"; "Fabric Design."

2 SPACE 2: Author(s)

General Instruction: After reading these instructions, decide who are the "authors" of this work for copyright purposes. Then, unless the work is a "collective work," give the requested information about every "author" who contributed any appreciable amount of copyrightable matter to this version of the work. If you need further space, request Continuation Sheets. In the case of a collective work, such as a catalog of paintings or collection of cartoons by various authors, give information about the author of the collective work as a whole.

Name of Author: The fullest form of the author's name should be given. Unless the work was "made for hire," the individual who actually created the work is its "author." In the case of a work made for hire, the statute provides that "the employer or other person for whom the work was prepared is considered the author."

What is a "Work Made for Hire"? A "work made for hire" is defined as: (1) "a work prepared by an employee within the scope of his or her employment"; or (2) "a work specially ordered or commissioned for use as a contribution to a collective work, as a part of a motion picture or other audiovisual work, as a translation, as a supplementary work, as a compilation, as an instructional text, as a test, as answer material for a test, or as an atlas, if the parties expressly agree in a written instrument signed by them that the work shall be considered a work made for hire." If you have checked "Yes" to indicate that the work was "made for hire," you must give the full legal name of the employer (or other person for whom the work was prepared). You may also include the name of the employee along with the name of the employer (for example: "Elster Publishing Co., employer for hire of John Ferguson").

"Anonymous" or "Pseudonymous" Work: An author's contribution to a work is "anonymous" if that author is not identified on the copies or phonorecords of the work. An author's contribution to a work is "pseudonymous" if that author is identified on the copies or phonorecords under a fictitious name. If the work is "anonymous" you may: (1) leave the line blank; or (2) state "anonymous" on the line; or (3) reveal the author's identity. If the work is "pseudonymous" you may: (1) leave the line blank; or (2) give the pseudonym and identify it as such (for example: "Huntley Haverstock, pseudonym"); or (3) reveal the author's name, making clear which is the real name and which is the pseudonym (for example: "Henry Leek, whose pseudonym is Priam Farrel"). However, the citizenship or domicile of the author must be given in all cases.

Dates of Birth and Death: If the author is dead, the statute requires that the year of death be included in the application unless the work is anonymous or pseudonymous. The author's birth date is optional but is useful as a form of identification. Leave this space blank if the author's contribution was a "work made for hire."

Author's Nationality or Domicile: Give the country of which the author is a citizen or the country in which the author is domiciled. Nationality or domicile must be given in all cases.

Nature of Authorship: Categories of pictorial, graphic, and sculptural authorship are listed below. Check the box(es) that best describe(s) each author's contribution to the work.

3-Dimensional sculptures: fine art sculptures, toys, dolls, scale models, and sculptural designs applied to useful articles.

2-Dimensional artwork: watercolor and oil paintings; pen and ink drawings; logo illustrations; greeting cards; collages; stencils; patterns; computer graphics; graphics appearing in screen displays; artwork appearing on posters, calendars, games, commercial prints and labels, and packaging, as well as 2-dimensional artwork applied to useful articles.

Reproductions of works of art: reproductions of preexisting artwork made by, for example, lithography, photoengraving, or etching.

Maps: cartographic representations of an area such as state and county maps, atlases, marine charts, relief maps, and globes.

Photographs: pictorial photographic prints and slides and holograms.

Jewelry designs: 3-dimensional designs applied to rings, pendants, earrings, necklaces, and the like.

Designs on sheetlike materials: designs reproduced on textiles, lace, and other fabrics; wallpaper; carpeting; floor tile; wrapping paper; and clothing.

Technical drawings: diagrams illustrating scientific or technical information in linear form such as architectural blueprints or mechanical drawings.

Text: textual material that accompanies pictorial, graphic, or sculptural works such as comic strips, greeting cards, games rules, commercial prints or labels, and maps.

Architectural works: designs of buildings, including the overall form as well as the arrangement and composition of spaces and elements of the design. NOTE: Any registration for the underlying architectural plans must be applied for on a separate Form VA, checking the box "Technical drawing."

3 SPACE 3: Creation and Publication

General Instructions: Do not confuse "creation" with "publication." Every application for copyright registration must state "the year in which creation of the work was completed." Give the date and nation of first publication only if the work has been published.

Creation: Under the statute, a work is "created" when it is fixed in a copy or phonorecord for the first time. Where a work has been prepared over a period of time, the part of the work existing in fixed form on a particular date constitutes the created work on that date. The date you give here should be the year in which the author completed the particular version for which registration is now being sought, even if other versions exist or if further changes or additions are planned.

Publication: The statute defines "publication" as "the distribution of copies or phonorecords of a work to the public by sale or other transfer of ownership, or by rental, lease, or lending"; a work is also "published" if there has been an "offering to distribute copies or phonorecords to a group of persons for purposes of further distribution, public performance, or public display." Give the full date (month, day, year) when, and the country where, publication first occurred. If first publication took place simultaneously in the United States and other countries, it is sufficient to state "U.S.A."

4 SPACE 4: Claimant(s)

Name(s) and Address(es) of Copyright Claimant(s): Give the name(s) and address(es) of the copyright claimant(s) in this work even if the claimant is the same as the author. Copyright in a work belongs initially to the author of the work (including, in the case of a work made for hire, the employer or other person for whom the work was prepared). The copyright claimant is either the author of the work or a person or organization to whom the copyright initially belonging to the author has been transferred.

Transfer: The statute provides that, if the copyright claimant is not the author, the application for registration must contain "a brief statement of how the claimant obtained ownership of the copyright." If any copyright claimant named in space 4 is not an author named in space 2, give a brief statement explaining how the claimant(s) obtained ownership of the copyright. Examples: "By written contract"; "Transfer of all rights by author"; "Assignment"; "By will." Do not attach transfer documents or other attachments or riders.

5 SPACE 5: Previous Registration

General Instructions: The questions in space 5 are intended to find out whether an earlier registration has been made for this work and, if so, whether

there is any basis for a new registration. As a rule, only one basic copyright registration can be made for the same version of a particular work.

Same Version: If this version is substantially the same as the work covered by a previous registration, a second registration is not generally possible unless: (1) the work has been registered in unpublished form and a second registration is now being sought to cover this first published edition; or (2) someone other than the author is identified as a copyright claimant in the earlier registration, and the author is now seeking registration in his or her own name. If either of these two exceptions apply, check the appropriate box and give the earlier registration number and date. Otherwise, do not submit Form VA; instead, write the Copyright Office for information about supplementary registration or recordation of transfers of copyright ownership.

Changed Version: If the work has been changed and you are now seeking registration to cover the additions or revisions, check the last box in space 5, give the earlier registration number and date, and complete both parts of space 6 in accordance with the instruction below.

Previous Registration Number and Date: If more than one previous registration has been made for the work, give the number and date of the latest registration.

6 SPACE 6: Derivative Work or Compilation

General Instructions: Complete space 6 if this work is a "changed version," "compilation," or "derivative work," and if it incorporates one or more earlier works that have already been published or registered for copyright, or that have fallen into the public domain. A "compilation" is defined as "a work formed by the collection and assembling of preexisting materials or of data that are selected, coordinated, or arranged in such a way that the resulting work as a whole constitutes an original work of authorship." A "derivative work" is "a work based on one or more preexisting works." Examples of derivative works include reproductions of works of art, sculptures based on drawings, lithographs based on paintings, maps based on previously published sources, or "any other form in which a work may be recast, transformed, or adapted." Derivative works also include works "consisting of editorial revisions, annotations, or other modifications" if these changes, as a whole, represent an original work of authorship.

Preexisting Material (space 6a): Complete this space and space 6b for derivative works. In this space identify the preexisting work that has been recast, transformed, or adapted. Examples of preexisting material might be "Grunewald Altarpiece" or "19th century quilt design." Do not complete this space for compilations.

Material Added to This Work (space 6b): Give a brief, general statement of the additional new material covered by the copyright claim for which registration is sought. In the case of a derivative work, identify this new material. Examples: "Adaptation of design and additional artistic work"; "Reproduction of painting by photolithography"; "Additional cartographic material"; "Compilation of photographs." If the work is a compilation, give a brief, general statement describing both the material that has been compiled and the compilation itself. Example: "Compilation of 19th century political cartoons."

7,8,9 SPACE 7,8,9: Fee, Correspondence, Certification, Return Address

Fee: The Copyright Office has the authority to adjust fees at 5-year intervals, based on changes in the Consumer Price Index. The next adjustment is due in 1996. Please contact the Copyright Office after July 1995 to determine the actual fee schedule.

Deposit Account: If you maintain a Deposit Account in the Copyright Office, identify it in space 7. Otherwise leave the space blank and send the fee of $20 with your application and deposit.

Correspondence (space 7): This space should contain the name, address, area code, and telephone number of the person to be consulted if correspondence about this application becomes necessary.

Certification (space 8): The application cannot be accepted unless it bears the date and the **handwritten signature** of the author or other copyright claimant, or of the owner of exclusive right(s), or of the duly authorized agent of the author, claimant, or owner of exclusive right(s).

Address for Return of Certificate (space 9): The address box must be completed legibly since the certificate will be returned in a window envelope.

FORM VA
For a Work of the Visual Arts
UNITED STATES COPYRIGHT OFFICE

REGISTRATION NUMBER

_____

VA   VAU

EFFECTIVE DATE OF REGISTRATION

_____

Month   Day   Year

DO NOT WRITE ABOVE THIS LINE. IF YOU NEED MORE SPACE, USE A SEPARATE CONTINUATION SHEET.

1   TITLE OF THIS WORK ▼   NATURE OF THIS WORK ▼ See instructions

PREVIOUS OR ALTERNATIVE TITLES ▼

PUBLICATION AS A CONTRIBUTION If this work was published as a contribution to a periodical, serial, or collection, give information about the collective work in which the contribution appeared.   **Title of Collective Work ▼**

If published in a periodical or serial give: Volume ▼   Number ▼   Issue Date ▼   On Pages ▼

---

2   **a**   NAME OF AUTHOR ▼   DATES OF BIRTH AND DEATH
                                                      Year Born ▼   Year Died ▼

Was this contribution to the work a "work made for hire"?   AUTHOR'S NATIONALITY OR DOMICILE
Name of Country   WAS THIS AUTHOR'S CONTRIBUTION TO THE WORK   If the answer to either of these questions is "Yes," see detailed instructions.

☐ Yes   OR {Citizen of ▶ _____
☐ No   {Domiciled in ▶ _____

Anonymous?   ☐ Yes   ☐ No
Pseudonymous?   ☐ Yes   ☐ No

NATURE OF AUTHORSHIP Check appropriate boxes(s). **See instructions**

☐ 3-Dimensional sculpture   ☐ Map   ☐ Technical drawing
☐ 2-Dimensional artwork   ☐ Photograph   ☐ Text
☐ Reproduction of work of art   ☐ Jewelry design   ☐ Architectural work
☐ Design on sheetlike material

NOTE

Under the law, the "author" of a "work made for hire" is generally the employer, not the employee (see instructions). For any part of this work that was "made for hire" check "Yes" in the space provided, give the employer (or other person for whom the work was prepared) as "Author" of that part, and leave the space for dates of birth and death blank.

b   NAME OF AUTHOR ▼   DATES OF BIRTH AND DEATH
                                               Year Born ▼   Year Died ▼

Was this contribution to the work a "work made for hire"?   AUTHOR'S NATIONALITY OR DOMICILE
Name of Country   WAS THIS AUTHOR'S CONTRIBUTION TO THE WORK   If the answer to either of these questions is "Yes," see detailed instructions.

☐ Yes   OR {Citizen of ▶ _____
☐ No   {Domiciled in ▶ _____

Anonymous?   ☐ Yes   ☐ No
Pseudonymous?   ☐ Yes   ☐ No

NATURE OF AUTHORSHIP Check appropriate boxes(s). **See instructions**

☐ 3-Dimensional sculpture   ☐ Map   ☐ Technical drawing
☐ 2-Dimensional artwork   ☐ Photograph   ☐ Text
☐ Reproduction of work of art   ☐ Jewelry design   ☐ Architectural work
☐ Design on sheetlike material

---

3   **a**   YEAR IN WHICH CREATION OF THIS WORK WAS COMPLETED   This information must be given ◀ Year in all cases.   **b**   DATE AND NATION OF FIRST PUBLICATION OF THIS PARTICULAR WORK
Complete this information ONLY if this work has been published.   Month ▶ _____ Day ▶ _____ Year ▶ _____   ◀ Nation

4   COPYRIGHT CLAIMANT(S) Name and address must be given even if the claimant is the same as the author given in space 2. ▼

See instructions before completing this space.

TRANSFER If the claimant(s) named here in space 4 is (are) different from the author(s) named in space 2, give a brief statement of how the claimant(s) obtained ownership of the copyright. ▼

APPLICATION RECEIVED

ONE DEPOSIT RECEIVED

TWO DEPOSITS RECEIVED

FUNDS RECEIVED

DO NOT WRITE HERE OFFICE USE ONLY

MORE ON BACK ▶   • Complete all applicable spaces (numbers 5-9) on the reverse side of this page.   **DO NOT WRITE HERE**
               • See detailed instructions.   • Sign the form at line 8.   Page 1 of _____ pages

EXAMINED BY	FORM VA
CHECKED BY	
☐ CORRESPONDENCE Yes	FOR COPYRIGHT OFFICE USE ONLY

DO NOT WRITE ABOVE THIS LINE. IF YOU NEED MORE SPACE, USE A SEPARATE CONTINUATION SHEET.

PREVIOUS REGISTRATION Has registration for this work, or for an earlier version of this work, already been made in the Copyright Office?
☐ Yes ☐ No If your answer is "Yes," why is another registration being sought? (Check appropriate box) ▼
a. ☐ This is the first published edition of a work previously registered in unpublished form.
b. ☐ This is the first application submitted by this author as copyright claimant.
c. ☐ This is a changed version of the work, as shown by space 6 on this application.
If your answer is "Yes," give: **Previous Registration Number** ▼   **Year of Registration** ▼

5

DERIVATIVE WORK OR COMPILATION Complete both space 6a and 6b for a derivative work; complete only 6b for a compilation.
a. **Preexisting Material** Identify any preexisting work or works that this work is based on or incorporates. ▼

b. **Material Added to This Work** Give a brief, general statement of the material that has been added to this work and in which copyright is claimed. ▼

See instructions before completing this space.

6

DEPOSIT ACCOUNT If the registration fee is to be charged to a Deposit Account established in the Copyright Office, give name and number of Account.
Name ▼   **Account Number** ▼

7

CORRESPONDENCE Give name and address to which correspondence about this application should be sent. Name/Address/Apt/City/State/ZIP ▼

Be sure to give your daytime phone ◄ number

Area Code and Telephone Number ▶

CERTIFICATION* I, the undersigned, hereby certify that I am the
check only one ▼
☐ author
☐ other copyright claimant
☐ owner of exclusive right(s)
☐ authorized agent of _____
     Name of author or other copyright claimant, or owner of exclusive right(s) ▲

of the work identified in this application and that the statements made
by me in this application are correct to the best of my knowledge.

8

Typed or printed name and date ▼ If this application gives a date of publication in space 3, do not sign and submit it before that date.
     Date ▶

☞ Handwritten signature (X) ▼

MAIL CERTIFI- CATE TO	Name ▼	**YOU MUST:** • Complete all necessary spaces • Sign your application in space 8	
	Number/Street/Apt ▼	**SEND ALL 3 ELEMENTS IN THE SAME PACKAGE:** 1. Application form 2. Nonrefundable $20 filing fee in check or money order payable to *Register of Copyrights* 3. Deposit material	The Copyright Office has the authority to ad- just fees at 5-year inter- vals, based on changes in the Consumer Price Index. The next adjust-
Certificate will be mailed in window envelope	City/State/ZIP ▼	**MAIL TO:** Register of Copyrights Library of Congress Washington, D.C. 20559-6000	ment is due in 1996. Please contact the Copyright Office after July 1995 to determine the actual fee schedule.

9

*17 U.S.C. § 506(e): Any person who knowingly makes a false representation of a material fact in the application for copyright registration provided for by section 409, or in any written statement filed in connection
with the application, shall be fined not more than $2,500.

July 1993—300,000   ♻ PRINTED ON RECYCLED PAPER   ☆U.S. GOVERNMENT PRINTING OFFICE: 1993-342-582/80,021

⊘Filling Out Short Form SE

BASIC INFORMATION

Read these instructions before completing this form.
Make sure all applicable spaces have been filled in
before you return this form.

When to Use This Form: All the following conditions must be met in order to use this form. If any one of the conditions does not apply, you must use Form SE. Incorrect use of this form will result in a delay in your registration.

The claim must be in a collective work.
The work must be essentially an all-new collective work or issue.
The author must be a citizen or domiciliary of the United States.
The work must be a work made for hire.
The author(s) and claimant(s) must be the same person(s) or organization(s).
The work must be first published in the United States.

Deposit to Accompany Application: An application for registration of a copyright claim in a serial issue first published in the United States must be accompanied by a deposit consisting of two copies (or phonorecords) of the best edition.

Fee: The filing fee of $20.00 must be sent for each issue to be registered. Do not send cash or currency.
  Copyright fees are adjusted at 5-year intervals, based on increases or decreases in the Consumer Price Index. The next adjustment is due in 1995. Contact the Copyright Office in January 1995 for the new fee schedule.

Mailing Requirements: It is important that you send the application, the deposit copies, and the $20.00 fee together in the same envelope or package. Send to: Register of Copyrights, Library of Congress, Washington, D.C. 20559.

Reproduction for Use of Blind or Physically Handicapped Individuals: A signature on this form and a check in one of these boxes constitutes a nonexclusive grant of permission to the Library of Congress to reproduce and distribute solely for the blind and physically handicapped under the conditions and limitations prescribed by the regulations of the Copyright Office: (1) copies of the work identified in space 1 of this application in Braille (or similar tactile symbols); or (2) phonorecords embodying a fixation of a reading of that work; or (3) both.

☐ Copies only   ☐ Phonorecords only   ☐ Copies and phonorecords

Collective Work: The term "collective work" refers to a work, such as a serial issue, in which a number of contributions are assembled into a collective whole. A claim in the "collective work" extends to all copyrightable authorship created by employees of the author, as well as any independent contributions in which the claimant has acquired ownership of the copyright.

Publication: The statute defines "publication" as "The distribution of copies or phonorecords of a work to the public by sale or other transfer of ownership, or by rental, lease, or lending;" a work is also "published" if there has been an "offering to distribute copies or phonorecords to a group of persons for purposes of further distribution, public performance, or public display."

Creation: A work is "created" when it is fixed in a copy (or phonorecord) for the first time.

Work Made for Hire: A "work made for hire" is defined as: (1) a work prepared by an employee within the scope of his or her employment; or (2) a work specially ordered or commissioned for certain uses (including use as a contribution to a collective work), if the parties expressly agree in a written instrument signed by them that the work shall be considered a work made for hire. The employer is the author of a work made for hire.

The Copyright Notice: For works first published on or after March 1, 1989, the law provides that a copyright notice in a specified form "may be placed on all publicly distributed copies from which the work can be visually perceived." Use of the copyright notice is the responsibility of the copyright owner and does not require advance permission from the Copyright Office. The required form of the notice for copies generally consists of three elements: (1) the symbol "©", or the word "Copyright," or the abbreviation "Copr."; (2) the year of first publication; and (3) the name of the owner of copyright. For example: "©1992 Jane Cole." The notice is to be affixed to the copies "in such manner and location as to give reasonable notice of the claim of copyright." Works first published prior to March 1, 1989, must carry the notice or risk loss of copyright protection.
  For information about notice requirements for works published before March 1, 1989, or other copyright information, write: Information Section, LM-401, Copyright Office, Library of Congress, Washington, D.C. 20559.

SPACE-BY-SPACE INSTRUCTIONS

1 SPACE 1: Title

  Every work submitted for copyright registration must be given a title to identify that particular work. Give the complete title of the periodical, including the volume, number, issue date, or other indicia printed on the copies. If possible, give the International Standard Serial Number (ISSN).

2 SPACE 2: Author and Copyright Claimant

  Give the fullest form of the author and claimant's name. If there are joint authors and owners, give the names of all the author/owners. (It is assumed that the authors and claimants are the same, that the work is made for hire, and that the claim is in the collective work).

3 SPACE 3: Date of Publication of This Particular Work

  Give the exact date on which publication of this issue first took place. The full date, including month, day, and year must be given.

Year in Which Creation of This Issue Was Completed: Give the year in which this serial issue was fixed in a copy or phonorecord for the first time. If no year is given, it is assumed that the issue was created in the same year in which it was published. The date must be the same as or no later than the publication date.

Certification: The application cannot be accepted unless it bears the handwritten signature of the copyright claimant or the duly authorized agent of the copyright claimant.

Person to Contact for Correspondence About This Claim: Give the name and telephone number, including area code, of the person to whom any correspondence concerning this claim should be addressed. Give the address only if it is different from the address for mailing of the certificate.

Deposit Account: If the filing fee is to be charged against a Deposit Account in the Copyright Office, give the name and number of the account in this space. Otherwise, leave the space blank and forward the $20.00 filing fee with your application and deposit.

Mailing Address of Certificate: This address must be complete and legible since the certificate will be mailed in a window envelope.

SHORT FORM SE

For a Serial
UNITED STATES COPYRIGHT OFFICE

REGISTRATION NUMBER

EFFECTIVE DATE OF REGISTRATION
(Assigned by Copyright Office)

Month	Day	Year

APPLICATION RECEIVED

ONE DEPOSIT RECEIVED

TWO DEPOSITS RECEIVED

EXAMINED BY

CORRESPONDENCE ☐

DO NOT WRITE ABOVE THIS LINE.

1 TITLE OF THIS SERIAL AS IT APPEARS ON THE COPY

Volume▼   Number▼   Date on Copies▼   ISSN▼

2 NAME AND ADDRESS OF THE AUTHOR AND COPYRIGHT CLAIMANT IN THIS COLLECTIVE WORK MADE FOR HIRE

3 DATE OF PUBLICATION OF THIS PARTICULAR ISSUE
Month▼   Day▼   Year▼

YEAR IN WHICH CREATION OF
THIS ISSUE WAS COMPLETED
(IF EARLIER THAN THE YEAR OF
PUBLICATION):
Year▼

CERTIFICATION*: I, the undersigned, hereby certify that I am the copyright claimant or the authorized agent of the copyright claimant of the work identified in this application, that all the conditions specified in the instructions on the back of this form are met, and that the statements made by me in this application are correct to the best of my knowledge.

Handwritten signature (X) _____

Typed or printed name of signer _____

PERSON TO CONTACT FOR CORRESPONDENCE ABOUT THIS CLAIM

Name ▶ _____
Daytime telephone number ▶ _____
Address (if other than given below) ▶ _____

DEPOSIT ACCOUNT

Account number ▶ _____
Name of account ▶ _____

MAIL
CERTIFI-
CATE TO

Name▼

Certificate
will be
mailed
in window
envelope

Number/Street/Apartment Number▼

City/State/ZIP▼

YOU MUST:
• Compare all necessary spaces
• Sign your application

**SEND ALL 3 ELEMENTS
IN THE SAME PACKAGE**
1. Application form
2. Nonrefundable $20 filing fee
   in check or money order
   payable to Register of Copyrights
3. Deposit material
MAIL TO
Register of Copyrights
Library of Congress
Washington, D.C. 20559

Copyright fees are adjusted at 5-year intervals, based on increases or decreases in the Consumer Price Index. The next adjustment is due in 1995. Contact the Copyright Office in January 1995 for the new fee schedule.

*17 U.S.C. §506(e): Any person who knowingly makes a false representation of a material fact in the application for copyright registration provided for by section 409, or in any written statement filed in connection with the application, shall be fined not more than $2,500.

June 1992—50,000

☆U.S. GOVERNMENT PRINTING OFFICE: 1992—312-432/60,001

✍ Filling Out Application Form SE

Detach and read these instructions before completing this form.
Make sure all applicable spaces have been filled in before you return this form.

BASIC INFORMATION

When To Use This Form: Use a separate Form SE for registration of each individual issue of a serial, Class SE. A serial is defined as a work issued or intended to be issued in successive parts bearing numerical or chronological designations and intended to be continued indefinitely. This class includes a variety of works; periodicals; newspapers; annuals; the journals, proceedings, transactions, etc., of societies. Do not use Form SE to register an individual contribution to a serial. Request Form TX for such contributions.

Deposit to Accompany Application: An application for copyright registration must be accompanied by a deposit consisting of copies or phonorecords representing the entire work for which registration is to be made. The following are the general deposit requirements as set forth in the statute:

> **Unpublished Work:** Deposit one complete copy (or phonorecord).

> **Published Work:** Deposit two complete copies (or phonorecord) of the best edition.

> **Work First Published Outside the United States:** Deposit one complete copy (or phonorecord) of the first foreign edition.

Mailing Requirements: It is important that you send the application, the deposit copy or copies, and the $20 fee together in the same envelope or package. The Copyright Office cannot process them unless they are received together. Send to: *Register of Copyrights, Library of Congress, Washington, D. C. 20559.*

The Copyright Notice: For works first published on or after March 1, 1989, the law provides that a copyright notice in a specified form "may be placed on all publicly distributed copies from which the work can be visually perceived." Use of the copyright notice is the responsibility of the copyright owner and does not require advance permission from the Copyright Office. The required form of the notice for copies generally consists of three elements: (1) the symbol "©," or the word "Copyright," or the abbreviation "Copr."; (2) the year of first publication; and (3) the name of the owner of copyright. For example: "© 1993 Jane Cole." The notice is to be affixed to the copies "in such manner and location as to give reasonable notice of the claim of copyright." Works first published prior to March 1, 1989, **must** carry the notice or risk loss of copyright protection.

For information about notice requirements for works published before March 1, 1989, or other copyright information, write: Information Section, LM-401, Copyright Office, Library of Congress, Washington, D.C. 20559.

LINE-BY-LINE INSTRUCTIONS

Please type or print using black ink.

---

1 SPACE 1: Title

Title of This Serial: Every work submitted for copyright registration must be given a title to identify that particular work. If the copies or phonorecords of the work bear a title (or an identifying phrase that could serve as a title), copy that wording *completely* and *exactly* on the application. Give the volume and number of the periodical issue for which you are seeking registration. The "Date on Copies" in space 1 should be the date appearing on the actual copies (for example: "June 1981," "Winter 1981"). Indexing of the registration and future identification of the work will depend on the information you give here.

Previous or Alternative Titles: Complete this space only if there are any additional titles for the serial under which someone searching for the registration might be likely to look or under which a document pertaining to the work might be recorded.

2 SPACE 2: Author(s)

General Instructions: After reading these instructions, decide who are the "authors" of this work for copyright purposes. In the case of a serial issue, an organization which directs the creation of the serial issue as a whole is generally considered the author of the "collective work" (see "Nature of Authorship") whether it employs a staff or uses the efforts of volunteers. Where, however, an individual is independently responsible for the serial issue, name that person as author of the "collective work."

Name of Author: The fullest form of the author's name should be given. In the case of a "work made for hire," the statute provides that "the employer or other person for whom the work was prepared is considered the author." If this issue is a "work made for hire," the author's name will be the full legal name of the hiring organization, corporation, or individual. The title of the periodical should not ordinarily be listed as "author" because the title itself does not usually correspond to a legal entity capable of authorship. When an individual creates an issue of a serial independently and not as an "employee" of an organization or corporation, that individual should be listed as the "author."

Author's Nationality or Domicile: Give the country of which the author is a citizen, or the country in which the author is domiciled. Nationality or domicile **must** be given in all cases. The citizenship of an organization formed under United States Federal or state law should be stated as "U.S.A."

What is a "Work Made for Hire"? A "work made for hire" is defined as (1) "a work prepared by an employee within the scope of his or her employment"; or (2) "a work specially ordered or commissioned for use as a contribution to a collective work, as a part of a motion picture or other audiovisual work, as a translation, as a supplementary work, as a compilation, as an instructional text, as a test, as answer material for a test, or as an atlas, if the parties expressly agree in a written instrument signed by them that the work shall be considered a work made for hire." An organization that uses the efforts of volunteers in the creation of a "collective work" (see "Nature of Authorship") may also be considered the author of a "work made for hire" even though those volunteers were not specifically paid by the organization. In the case of a "work made for hire," give the full legal name of the employer and check "Yes" to indicate that the work was made for hire. You may also include the name of the employee along with the name of the employer (for example: "Elster Publishing Co., employer for hire of John Ferguson").

"Anonymous" or "Pseudonymous" Work: Leave this space **blank** if the serial is a "work made for hire." An author's contribution to a work is "anonymous" if that author is not identified on the copies or phonorecords of the work. An author's contribution to a work is "pseudonymous" if that author is identified on the copies or phonorecords under a fictitious name. If the work is "anonymous" you may: (1) leave the line blank; or (2) state "anonymous" on the line; or (3) reveal the author's identity. If the work is "pseudonymous" you may: (1) leave the line blank; or (2) give the pseudonym and identify it as such (for example: "Huntley Haverstock, pseudonym"); or (3) reveal the author's name, making clear which is the real name and which is the pseudonym (for example: "Judith Barton, whose pseudonym is Madeline Elster"). However, the citizenship or domicile of the author **must** be given in all cases.

Dates of Birth and Death: Leave this space blank if the author's contribution was a "work made for hire." If the author is dead, the statute requires that the year of death be included in the application unless the work is anonymous or pseudonymous. The author's birth date is optional but is useful as a form of identification.

Nature of Authorship: Give a brief statement of the nature of the particular author's contribution to the work. If an organization directed, controlled, and supervised the creation of the serial issue as a whole, check the box "collective work." The term "collective work" means that the author is responsible for compilation and editorial revision and may also be responsible for certain individual contributions to the serial issue. Further examples of "Authorship" which may apply both to organizational and to individual authors are "Entire text"; "Entire text and/or illustrations"; "Editorial revision, compilation, plus additional new material."

3 SPACE 3: Creation and Publication

General Instructions: Do not confuse "creation" with "publication." Every application for copyright registration must state "the year in which creation of the work was completed." Give the date and nation of first publication only if the work has been published.

Creation: Under the statute, a work is "created" when it is fixed in a copy or phonorecord for the first time. Where a work has been prepared over a period of time, the part of the work existing in fixed form on a particular date constitutes the created work on that date. The date you give here should be the year in which this particular issue was completed.

Publication: The statute defines "publication" as "the distribution of copies or phonorecords of a work to the public by sale or other transfer of ownership or by rental, lease, or lending"; a work is also "published" if there has been an "offering to distribute copies or phonorecords to a group of persons for purposes of further distribution, public performance, or public display." Give the full date (month, day, year) when, and the country where, publication of this particular issue first occurred. If first publication took place simultaneously in the United States and other countries, it is sufficient to state "U.S.A."

4 SPACE 4: Claimant(s)

Name(s) and Address(es) of Copyright Claimant(s): This space must be completed. Give the name(s) and address(es) of the copyright claimant(s) of this work even if the claimant is the same as the author named in space 2. Copyright in a work belongs initially to the author of the work (including, in the case of a work made for hire, the employer or other person for whom the work was prepared). The copyright claimant is either the author of the work or a person or organization to whom the copyright initially belonging to the author has been transferred.

Transfer: The statute provides that, if the copyright claimant is not the author, the application for registration must contain "a brief statement of how the claimant obtained ownership of the copyright." If any copyright claimant named in space 4 is not an author named in space 2, give a brief statement explaining how the claimant(s) obtained ownership of the copyright. Examples: "By written contract"; "Transfer of all rights by author"; "Assignment"; "By will." Do not attach transfer documents or other attachments or riders.

5 SPACE 5: Previous Registration

General Instructions: This space rarely applies to serials. Complete space 5 if this particular issue has been registered earlier or if it contains a substantial amount of material that has been previously registered. Do not complete this space if the previous registrations are simply those made for earlier issues.

Previous Registration:
a. Check this box if this issue has been registered in unpublished form and a second registration is now sought to cover the first published edition.
b. Check this box if someone other than the author is identified as copyright claimant in the earlier registration and the author is now seeking registration in his or her own name. If the work in question is a contribution to a collective work as opposed to the issue as a whole, file Form TX, not Form SE.
c. Check this box (and complete space 6) if this particular issue or a substantial portion of the material in it has been previously registered and you are now seeking registration for the additions and revisions which appear in this issue for the first time.

Previous Registration Number and Date: Complete this line if you checked one of the boxes above. If more than one previous registration has been made for the issue or for material in it, give only the number and year date for the latest registration.

6 SPACE 6: Derivative Work or Compilation

General Instructions: Complete space 6 if this issue is a "changed version," "compilation," or "derivative work" that incorporates one or more earlier works that have already been published or registered for copyright or that have fallen into the public domain. Do not complete space 6 for an issue consisting of entirely new material appearing for the first time such as a new issue of a continuing serial. A "compilation" is defined as "a work formed by the

collection and assembling of preexisting materials or of data that are selected, coordinated, or arranged in such a way that the resulting work as a whole constitutes an original work of authorship." A "derivative work" is "a work based on one or more preexisting works." Examples of derivative works include translations, fictionalizations, abridgments, condensations, or "any other form in which a work may be recast, transformed, or adapted." Derivative works also include works "consisting of editorial revisions, annotations, or other modifications" if these changes, as a whole, represent an original work of authorship.

Preexisting Material (space 6a): For derivative works, complete this space and space 6b. In space 6a identify the preexisting work that has been recast, transformed, adapted, or updated. Example: "1978 Morgan Co. Sales Catalog." Do not complete space 6a for compilations.

Material Added to This Work (space 6b): Give a brief, general statement of the new material covered by the copyright claim for which registration is sought. Examples include: "Editorial revisions and additions to the Catalog"; "Translation"; "Additional material." If a periodical issue is a compilation, describe both the compilation itself and the material that has been compiled. Examples: "Compilation of previously published journal articles"; "Compilation of previously published data." An issue may be both a derivative work and a compilation, in which case a sample statement might be: "Compilation of [describe] and additional new material."

7 SPACE 7: Manufacturing Provisions

Due to the expiration of the Manufacturing Clause of the copyright law on June 30, 1986, this space has been deleted.

8 SPACE 8: Reproduction for Use of Blind or Physically Handicapped Individuals

General Instructions: One of the major programs of the Library of Congress is to provide Braille editions and special recordings of works for the exclusive use of the blind and physically handicapped. In an effort to simplify and speed up the copyright licensing procedures that are a necessary part of this program, section 710 of the copyright statute provides for the establishment of a voluntary licensing system to be tied in with copyright registration. Copyright Office regulations provide that you may grant a license for such reproduction and distribution solely for the use of persons who are certified by competent authority as unable to read normal printed material as a result of physical limitations. The license is entirely voluntary, nonexclusive, and may be terminated upon 90 days notice.

How to Grant the License: If you wish to grant it, check one of the three boxes in space 8. Your check in one of these boxes together with your signature in space 10 will mean that the Library of Congress can proceed to reproduce and distribute under the license without further paperwork. For further information, write for Circular 63.

9,10,11 SPACE 9,10,11: Fee, Correspondence, Certification, Return Address

Fee: The Copyright Office has the authority to adjust fees at 5-year intervals, based on changes in the Consumer Price Index. The next adjustment is due in 1996. Please contact the Copyright Office after July 1995 to determine the actual fee schedule.
Deposit Account: If you maintain a Deposit Account in the Copyright Office, identify it in space 9. Otherwise leave the space blank and send the fee of $20 with your application and deposit.

Correspondence (space 9): This space should contain the name, address, area code, and telephone number of the person to be consulted if correspondence about this application becomes necessary.

Certification (space 10): The application cannot be accepted unless it bears the date and the **handwritten signature** of the author or other copyright claimant, or of the owner of exclusive right(s), or of the duly authorized agent of the author, claimant, or owner of exclusive right(s).

Address for Return of Certificate (space 11): The address box must be completed legibly since the certificate will be returned in a window envelope.

FORM SE

For a Serial
UNITED STATES COPYRIGHT OFFICE

REGISTRATION NUMBER

_____

U

EFFECTIVE DATE OF REGISTRATION

_____ _____ _____
Month   Day   Year

DO NOT WRITE ABOVE THIS LINE. IF YOU NEED MORE SPACE, USE A SEPARATE CONTINUATION SHEET.

1

TITLE OF THIS SERIAL ▼

Volume ▼   Number ▼   Date on Copies ▼   Frequency of Publication ▼

PREVIOUS OR ALTERNATIVE TITLES ▼

2

a

NAME OF AUTHOR ▼

DATES OF BIRTH AND DEATH
Year Born ▼   Year Died ▼

Was this contribution to the work a "work made for hire"?
☐ Yes
☐ No

AUTHOR'S NATIONALITY OR DOMICILE
Name of Country
OR { Citizen of ▶ _____
      Domiciled in▶ _____

WAS THIS AUTHOR'S CONTRIBUTION TO THE WORK
Anonymous?   ☐ Yes ☐ No
Pseudonymous?   ☐ Yes ☐ No
If the answer to either of these questions is "Yes," see detailed instructions.

NATURE OF AUTHORSHIP Briefly describe nature of material created by this author in which copyright is claimed. ▼
☐ Collective Work   Other:

NOTE

Under the law, the "author" of a "work made for hire" is generally the employer, not the employee (see instructions). For any part of this work that was "made for hire" check "Yes" in the space provided, give the employer (or other person for whom the work was prepared) as "Author" of that part, and leave the space for dates of birth and death blank.

b

NAME OF AUTHOR ▼

DATES OF BIRTH AND DEATH
Year Born ▼   Year Died ▼

Was this contribution to the work a "work made for hire"?
☐ Yes
☐ No

AUTHOR'S NATIONALITY OR DOMICILE
Name of Country
OR { Citizen of ▶ _____
      Domiciled in▶ _____

WAS THIS AUTHOR'S CONTRIBUTION TO THE WORK
Anonymous?   ☐ Yes ☐ No
Pseudonymous?   ☐ Yes ☐ No
If the answer to either of these questions is "Yes," see detailed instructions.

NATURE OF AUTHORSHIP Briefly describe nature of material created by this author in which copyright is claimed. ▼
☐ Collective Work   Other:

c

NAME OF AUTHOR ▼

DATES OF BIRTH AND DEATH
Year Born ▼   Year Died ▼

Was this contribution to the work a "work made for hire"?
☐ Yes
☐ No

AUTHOR'S NATIONALITY OR DOMICILE
Name of Country
OR { Citizen of ▶ _____
      Domiciled in▶ _____

WAS THIS AUTHOR'S CONTRIBUTION TO THE WORK
Anonymous?   ☐ Yes ☐ No
Pseudonymous?   ☐ Yes ☐ No
If the answer to either of these questions is "Yes," see detailed instructions.

NATURE OF AUTHORSHIP Briefly describe nature of material created by this author in which copyright is claimed. ▼
☐ Collective Work   Other:

3

a

YEAR IN WHICH CREATION OF THIS ISSUE WAS COMPLETED
This information must be given
◀ Year in all cases.

b

DATE AND NATION OF FIRST PUBLICATION OF THIS PARTICULAR ISSUE
Complete this information Month ▶ _____ Day▶ _____ Year▶ _____
ONLY if this work has been published.
◀ Nation

4

See instructions before completing this space

COPYRIGHT CLAIMANT(S) Name and address must be given even if the claimant is the same as the author given in space 2. ▼

TRANSFER If the claimant(s) named here in space 4 is (are) different from the author(s) named in space 2, give a brief statement of how the claimant(s) obtained ownership of the copyright. ▼

APPLICATION RECEIVED

ONE DEPOSIT RECEIVED

TWO DEPOSITS RECEIVED

REMITTANCE NUMBER AND DATE

DO NOT WRITE HERE
OFFICE USE ONLY

MORE ON BACK ▶
• Complete all applicable spaces (numbers 5-11) on the reverse side of this page.
• See detailed instructions.   • Sign the form at line 10.

DO NOT WRITE HERE
Page 1 of _____ pages

EXAMINED BY	FORM SE
CHECKED BY	
☐ CORRESPONDENCE Yes	FOR COPYRIGHT OFFICE USE ONLY

DO NOT WRITE ABOVE THIS LINE. IF YOU NEED MORE SPACE, USE A SEPARATE CONTINUATION SHEET.

PREVIOUS REGISTRATION Has registration for this issue, or for an earlier version of this particular issue, already been made in the Copyright Office?
☐ Yes ☐ No If your answer is "Yes," why is another registration being sought? (Check appropriate box) ▼
a. ☐ This is the first published edition of an issue previously registered in unpublished form.
b. ☐ This is the first application submitted by this author as copyright claimant.
c. ☐ This is a changed version of this issue, as shown by space 6 on this application.
If your answer is "Yes," give: **Previous Registration Number** ▼   **Year of Registration** ▼

5

DERIVATIVE WORK OR COMPILATION Complete both space 6a and 6b for a derivative work; complete only 6b for a compilation.
a. **Preexisting Material** Identify any preexisting work or works that this work is based on or incorporates. ▼

b. **Material Added to This Work** Give a brief, general statement of the material that has been added to this work and in which copyright is claimed. ▼

6

See instructions before completing this space.

—space deleted—

7

REPRODUCTION FOR USE OF BLIND OR PHYSICALLY HANDICAPPED INDIVIDUALS A signature on this form at space 10 and a check in one of the boxes here in space 8 constitutes a non-exclusive grant of permission to the Library of Congress to reproduce and distribute solely for the blind and physically handicapped and under the conditions and limitations prescribed by the regulations of the Copyright Office: (1) copies of the work identified in space 1 of this application in Braille (or similar tactile symbols); or (2) phonorecords embodying a fixation of a reading of that work; or (3) both.

a ☐ Copies and Phonorecords   b ☐ Copies Only   c ☐ Phonorecords Only

8

See instructions.

DEPOSIT ACCOUNT If the registration fee is to be charged to a Deposit Account established in the Copyright Office, give name and number of Account.
Name ▼   **Account Number** ▼

CORRESPONDENCE Give name and address to which correspondence about this application should be sent.   Name/Address/Apt/City/State/ZIP ▼

Area Code and Telephone Number ▶

9

Be sure to give your daytime phone number ◀

CERTIFICATION* I, the undersigned, hereby certify that I am the
Check only one ▶
☐ author
☐ other copyright claimant
☐ owner of exclusive right(s)
☐ authorized agent of _____
of the work identified in this application and that the statements made by me in this application are correct to the best of my knowledge.
Name of author or other copyright claimant, or owner of exclusive right(s) ▲

Typed or printed name and date ▼ If this application gives a date of publication in space 3, do not sign and submit it before that date.

date ▶ _____

☞ Handwritten signature (X) ▼

10

MAIL CERTIFI-CATE TO
Certificate will be mailed in window envelope

Name ▼

Number/Street/Apartment Number ▼

City/State/ZIP ▼

YOU MUST:
• Complete all necessary spaces
• Sign your application in space 10

SEND ALL 3 ELEMENTS IN THE SAME PACKAGE:
1. Application form
2. Nonrefundable $20 filing fee in check or money order payable to *Register of Copyrights*
3. Deposit material

MAIL TO:
Register of Copyrights
Library of Congress
Washington, D.C. 20559-6000

The Copyright Office has the authority to adjust fees at 5-year intervals, based on changes in the Consumer Price Index. The next adjustment is due in 1996. Please contact the Copyright Office after July 1995 to determine the actual fee schedule.

11

*17 U.S.C. § 506(e): Any person who knowingly makes a false representation of a material fact in the application for copyright registration provided for by section 409, or in any written statement filed in connection with the application, shall be fined not more than $2,500.

April 1993—100,000   ☆U.S. GOVERNMENT PRINTING OFFICE: 1993-342-581/60,511

⊕Form SE/GROUP

BASIC INFORMATION

Read these instructions before completing this form.
Make sure all applicable spaces have been filled in
before you return this form.

When to Use This Form:
All the following conditions must be met in order to use this form. If any one of the conditions does not apply, you must register the issues separately using Form SE or Short Form SE.

1. You must have given a complimentary subscription for two copies of the serial to the Library of Congress, confirmed by letter to the GeneralCounsel, Copyright Office. Subscription copies must be mailed **separately** to:
   Library of Congress
   Group Periodicals Registration
   Washington, D.C. 20540
2. The claim must be in the collective works.
3. The works must be essentially all new collective works or issues.
4. Each issue must be a work made for hire.
5. The author(s) and claimant(s) must be the same person(s) or organization(s) for all of the issues.
6. Each issue must have been created no more than one year prior to publication.
7. All issues in the group must have been published within the same calendar year.

Which Issues May Be Included in a Group Registration:
You may register two or more issues of a serial published at intervals of one week or longer under the same continuing title, provided that the issues were published within a 90-day period during the same calendar year.

Deposit to Accompany Application:
Send one copy of each issue included in the group registration with the application and fee.

Fee:
A nonrefundable filing fee of $10.00 FOR EACH ISSUE LISTED ON FORM SE/GROUP must be sent with the application or charged to an active deposit account in the Copyright Office. There is a minimum fee of $20.00 for Form SE/Group. Special handling is not available for Form SE/Group.

Mailing Instructions:
Send the application, deposit copies, and fee together in the same package to: Register of Copyrights, Library of Congress, Washington, D.C. 20559.

International Standard Serial Number (ISSN):
ISSN is an internationally accepted code for the identification of serial publications. If a published serial has not been assigned an ISSN, application forms and additional information may be obtained from National Serials Data Program, Library of Congress, Washington, D.C. 20540. Do not contact the Copyright Office for ISSNs.

Collective Work:
The term "collective work" refers to a work, such as a serial issue, in which a number of contributions are assembled into a collective whole. A claim in the "collective work" extends to all copyrightable authorship created by employees of the author, as well as any independent contributions in which the claimant has acquired ownership of the copyright.

Publication:
The statute defines "publication" as "The distribution of copies or phonorecords of a work to the public by sale or other transfer of ownership, or by rental, lease, or lending;" a work is also "published" if there has been an "offering to distribute copies or phonorecords to a group of persons for purposes of further distribution, public performance, or public display."

Creation:
A work is "created" when it is fixed in a copy (or phonorecord) for the first time. For a serial, the year in which the collective work was completed is the creation date.

Work Made for Hire:
A "work made for hire" is defined as: (1) a work prepared by an employee within the scope of his or her employment; or (2) a work specially ordered or commissioned for certain uses (including use as a contribution to a collective work), if the parties expressly agree in a written instrument signed by them that the work shall be considered a work made for hire. The employer is the author of a work made for hire.

The Copyright Notice:
For works first published on or after March 1, 1989, the law provides that a copyright notice in a specified form "may be placed on all publicly distributed copies from which the work can be visually perceived." Use of the copyright notice is the responsibility of the copyright owner and does not require advance permission from the Copyright Office. The required form of the notice for copies generally consists of three elements: (1) the symbol "©", or the word "Copyright," or the abbreviation "Copr."; (2) the year of first publication; and (3) the name of the owner of copyright. For example: "©1990 Jane Cole." The notice is to be affixed to the copies "in such manner and location as to give reasonable notice of the claim of copyright." Works first published prior to March 1, 1989, **must** carry the notice or risk loss of copyright protection.

For information about notice requirements for works published before March 1, 1989, or other copyright information, write: Information Section, LM-401, Copyright Office, Library of Congress, Washington, D.C. 20559.

PRIVACY ACT ADVISORY STATEMENT Required by the Privacy Act of 1974 (P.L. 93-579)
The authority for requesting this information is title 17 U.S.C. secs. 409 and 410. Furnishing the requested information is voluntary. But if the information is not furnished, it may be necessary to delay or refuse registration and you may not be entitled to certain relief, remedies, and benefits provided in chapters 4 and 5 of title 17, U.S.C.

The principal uses of the requested information are the establishment and maintenance of a public record and the examination of the application for compliance with legal requirements. Other routine uses include public inspection and copying, preparation of public indexes, preparation of public catalogs of copyright registrations, and preparation of search reports upon request.

NOTE: No other advisory statement will be given in connection with this application. Please keep this statement and refer to it if we communicate with you regarding this application.

SPACE-BY-SPACE INSTRUCTIONS

1 SPACE 1: Title and Date of Publication

Give the complete title of the serial, followed by the International Standard Serial Number (ISSN), if available. List the issues in the order of publication. For each issue, give the volume, number, and issue date appearing on the copies, followed by the complete date of publication, including month, day, and year. If you have not previously registered this **identical** title under Section 408 of the Copyright Act, please indicate by checking the box.

2 SPACE 2: Author and Copyright Claimant

Give the fullest form of the author and claimant's name and mailing address. If there are joint authors and claimants, give the names and addresses of all the author/claimants. If the work is not of U.S. origin, add the citizenship or domicile of the author/claimant, or the nation of publication.

Certification:
The application cannot be accepted unless it bears the handwritten signature of the copyright claimant or the duly authorized agent of the copyright claimant.

Person to Contact for Correspondence About This Claim:
Give the name and telephone number, including area code, of the person to whom any correspondence concerning this claim should be addressed. Give the address only if it is different from the address for mailing of the certificate.

Deposit Account:
If the filing fee is to be charged against a deposit account in the Copyright Office, give the name and number of the account in this space. Otherwise, leave the space blank and forward the filing fee with your application and deposit.

Mailing Address of Certificate:
This address must be complete and legible since the certificate will be mailed in a window envelope.

Reproduction for Use of Blind or Physically Handicapped Individuals:
A signature on this form and a check in one of these boxes constitutes a nonexclusive grant of permission to the Library of Congress to reproduce and distribute solely for the blind and physically handicapped under the conditions and limitations prescribed by the regulations of the Copyright Office: (1) copies of the work identified in space 1 of this application in Braille (or similar tactile symbols); or (2) phonorecords embodying a fixation of a reading of that work; or (3) both.

FORM SE/GROUP
UNITED STATES COPYRIGHT OFFICE

REGISTRATION NUMBER

EFFECTIVE DATE OF REGISTRATION
(Assigned by Copyright Office)

Month	Day	Year

APPLICATION RECEIVED

ONE DEPOSIT RECEIVED

EXAMINED BY	CORRESPONDENCE ☐

DO NOT WRITE ABOVE THIS LINE.

1

TITLE ▼   ISSN ▼

List in order of publication

No previous registration under identical title ☐

	Volume ▼	Number ▼	Issue date on copies ▼	Month, day and year of publication ▼
1.				
2.				
3.				
4.				
5.				
6.				
7.				
8.				
9.				
10.				
11.				
12.				
13.				
14.				

2

NAME AND ADDRESS OF THE AUTHOR/COPYRIGHT CLAIMANT IN THESE COLLECTIVE WORKS MADE FOR HIRE

FOR NON-U.S. WORKS: Author's citizenship ▼   Domicile ▼   Nation of publication ▼

CERTIFICATION*: I, the undersigned, hereby certify that I am the copyright claimant or the authorized agent of the copyright claimant of the works identified in this application, that all the conditions specified in the instructions on the back of this form are met, that I have deposited two complimentary subscription copies with the Library of Congress, and that the statements made by me in this application are correct to the best of my knowledge.

Signature (X) _____   Typed or printed name _____

PERSON TO CONTACT FOR CORRESPONDENCE ABOUT THIS CLAIM

Name ▶ _____

Daytime telephone number ▶ _____

Address (if other than given below) ▶ _____

DEPOSIT ACCOUNT

Account number ▶ _____

Name of account ▶ _____

MAIL CERTIFI-CATE TO

Certificate will be mailed in window envelope

Name ▼

Number/Street/Apartment Number ▼

City/State/ZIP ▼

REPRODUCTION FOR USE OF BLIND OR PHYSICALLY HANDICAPPED INDIVIDUALS

a ☐ Copies and Phonorecords

b ☐ Copies Only

c ☐ Phonorecords Only

MAIL TO:
Register of Copyrights
Library of Congress
Washington, D.C. 20559

*17 U.S.C. §506(e): Any person who knowingly makes a false representation of a material fact in the application for copyright registration provided for by section 409, or in any written statement filed in connection with the application, shall be fined not more than $2,500.

April 1991—100,000

☆ U.S. GOVERNMENT PRINTING OFFICE: 1991-282-170.20.016

⊘Filling Out Application Form PA

Detach and read these instructions before completing this form. Make sure all applicable spaces have been filled in before you return this form.

BASIC INFORMATION

When to Use This Form: Use Form PA for registration of published or unpublished works of the performing arts. This class includes works prepared for the purpose of being "performed" directly before an audience or indirectly "by means of any device or process." Works of the performing arts include: (1) musical works, including any accompanying words; (2) dramatic works, including any accompanying music; (3) pantomimes and choreographic works; and (4) motion pictures and other audiovisual works.

Deposit to Accompany Application: An application for copyright registration must be accompanied by a deposit consisting of copies or phonorecords representing the entire work for which registration is to be made. The following are the general deposit requirements as set forth in the statute:

Unpublished Work: Deposit one complete copy (or phonorecord).

Published Work: Deposit two complete copies or one phonorecord of the best edition.

Work First Published Outside the United States: Deposit one complete copy (or phonorecord) of the first foreign edition.

Contribution to a Collective Work: Deposit one complete copy (or phonorecord) of the best edition of the collective work.

Motion Pictures: Deposit *both* of the following: (1) a separate written description of the contents of the motion picture; and (2) for a published work, one complete copy of the best edition of the motion picture; or, for an unpublished work, one complete copy of the motion picture or identifying material. Identifying material may be either an audiorecording of the entire soundtrack or one frame enlargement or similar visual print from each 10-minute segment.

The Copyright Notice: For works first published on or after March 1, 1989, the law provides that a copyright notice in a specified form "may be placed on all publicly distributed copies from which the work can be visually perceived." Use of the copyright notice is the responsibility of the copyright owner and does not require advance permission from the Copyright Office. The required form of the notice for copies generally consists of three elements: (1) the symbol "©", or the word "Copyright," or the abbreviation "Copr."; (2) the year of first publication; and (3) the name of the owner of copyright. For example: "© 1989 Jane Cole." The notice is to be affixed to the copies "in such manner and location as to give reasonable notice of the claim of copyright." Works first published prior to March 1, 1989, must carry the notice or risk loss of copyright protection.

For information about notice requirements for works published before March 1, 1989, or other copyright information, write: Information Section, LM-401, Copyright Office, Library of Congress, Washington, D.C. 20559.

LINE-BY-LINE INSTRUCTIONS

1 SPACE 1: Title

Title of This Work: Every work submitted for copyright registration must be given a title to identify that particular work. If the copies or phonorecords of the work bear a title (or an identifying phrase that could serve as a title), transcribe that wording *completely* and *exactly* on the application. Indexing of the registration and future identification of the work will depend on the information you give here. If the work you are registering is an entire "collective work" (such as a collection of plays or songs), give the overall title of the collection. If you are registering one or more individual contributions to a collective work, give the title of each contribution, followed by the title of the collective work. Example: "'A Song for Elinda' in *Old and New Ballads for Old and New People*."

Previous or Alternative Titles: Complete this space if there are any additional titles for the work under which someone searching for the registration might be likely to look, or under which a document pertaining to the work might be recorded.

Nature of This Work: Briefly describe the general nature or character of the work being registered for copyright. Examples: "Music"; "Song Lyrics"; "Words and Music"; "Drama"; "Musical Play"; "Choreography"; "Pantomime"; "Motion Picture"; "Audiovisual Work."

2 SPACE 2: Author(s)

General Instructions: After reading these instructions, decide who are the "authors" of this work for copyright purposes. Then, unless the work is a "collective work," give the requested information about every "author" who contributed any appreciable amount of copyrightable matter to this version of the work. If you need further space, request additional Continuation Sheets. In the case of a collective work, such as a songbook or a collection of plays, give information about the author of the collective work as a whole.

Name of Author: The fullest form of the author's name should be given. Unless the work was "made for hire," the individual who actually created the work is its "author." In the case of a work made for hire, the statute provides

that "the employer or other person for whom the work was prepared is considered the author."

What is a "Work Made for Hire"? A "work made for hire" is defined as: (1) "a work prepared by an employee within the scope of his or her employment"; or (2) "a work specially ordered or commissioned for use as a contribution to a collective work, as a part of a motion picture or other audiovisual work, as a translation, as a supplementary work, as a compilation, as an instructional text, as a test, as answer material for a test, or as an atlas, if the parties expressly agree in a written instrument signed by them that the work shall be considered a work made for hire." If you have checked "Yes" to indicate that the work was "made for hire," you must give the full legal name of the employer (or other person for whom the work was prepared). You may also include the name of the employee along with the name of the employer (for example: "Elster Music Co., employer for hire of John Ferguson").

"Anonymous" or "Pseudonymous" Work: An author's contribution to a work is "anonymous" if that author is not identified on the copies or phonorecords of the work. An author's contribution to a work is "pseudonymous" if that author is identified on the copies or phonorecords under a fictitious name. If the work is "anonymous" you may: (1) leave the line blank; or (2) state "anonymous" on the line; or (3) reveal the author's identity. If the work is "pseudonymous" you may: (1) leave the line blank; or (2) give the pseudonym and identify it as such (for example: "Huntley Haverstock, pseudonym"); or (3) reveal the author's name, making clear which is the real name and which the pseudonym (for example: "Judith Barton, whose pseudonym is Madeline Elster"). However, the citizenship or domicile of the author must be given in all cases.

Dates of Birth and Death: If the author is dead, the statute requires that the year of death be included in the application unless the work is anonymous or pseudonymous. The author's birth date is optional, but is useful as a form of identification. Leave this space blank if the author's contribution was a "work made for hire."

Author's Nationality or Domicile: Give the country of which the author is a citizen, or the country in which the author is domiciled. Nationality or domicile must be given in all cases.

Nature of Authorship: Give a brief general statement of the nature of this particular author's contribution to the work. Examples: "Words"; "Co-Author of Music"; "Words and Music"; "Arrangement"; "Co-Author of Book and Lyrics"; "Dramatization"; "Screen Play"; "Compilation and English Translation"; "Editorial Revisions."

3 SPACE 3: Creation and Publication

General Instructions: Do not confuse "creation" with "publication." Every application for copyright registration must state "the year in which creation of the work was completed." Give the date and nation of first publication only if the work has been published.

Creation: Under the statute, a work is "created" when it is fixed in a copy or phonorecord for the first time. Where a work has been prepared over a period of time, the part of the work existing in fixed form on a particular date constitutes the created work on that date. The date you give here should be the year in which the author completed the particular version for which registration is now being sought, even if other versions exist or if further changes or additions are planned.

Publication: The statute defines "publication" as "the distribution of copies or phonorecords of a work to the public by sale or other transfer of ownership, or by rental, lease, or lending"; a work is also "published" if there has been an "offering to distribute copies or phonorecords to a group of persons for purposes of further distribution, public performance, or public display." Give the full date (month, day, year) when, and the country where, publication first occurred. If first publication took place simultaneously in the United States and other countries, it is sufficient to state "U.S.A."

4 SPACE 4: Claimant(s)

Name(s) and Address(es) of Copyright Claimant(s): Give the name(s) and address(es) of the copyright claimant(s) in this work even if the claimant is the same as the author. Copyright in a work belongs initially to the author of the work (including, in the case of a work made for hire, the employer or other person for whom the work was prepared). The copyright claimant is either the author of the work or a person or organization to whom the copyright initially belonging to the author has been transferred.

Transfer: The statute provides that, if the copyright claimant is not the author, the application for registration must contain "a brief statement of how the claimant obtained ownership of the copyright." If any copyright claimant named in space 4 is not an author named in space 2, give a brief statement explaining how the claimant(s) obtained ownership of the copyright. Examples: "By written contract"; "Transfer of all rights by author"; "Assignment"; "By will." Do not attach transfer documents or other attachments or riders.

5 SPACE 5: Previous Registration

General Instructions: The questions in space 5 are intended to find out whether an earlier registration has been made for this work and, if so, whether there is any basis for a new registration. As a general rule, only one basic copyright registration can be made for the same version of a particular work.

Same Version: If this version is substantially the same as the work covered by a previous registration, a second registration is not generally possible unless: (1) the work has been registered in unpublished form and a second registration is now being sought to cover this first published edition; or (2) someone other than the author is identified as copyright claimant in the earlier registration, and the author is now seeking registration in his or her own name. If either of these two exceptions apply, check the appropriate box and give the

earlier registration number and date. Otherwise, do not submit Form PA; instead, write the Copyright Office for information about supplementary registration or recordation of transfers of copyright ownership.

Changed Version: If the work has been changed, and you are now seeking registration to cover the additions or revisions, check the last box in space 5, give the earlier registration number and date, and complete both parts of space 6 in accordance with the instructions below.

Previous Registration Number and Date: If more than one previous registration has been made for the work, give the number and date of the latest registration.

6 SPACE 6: Derivative Work or Compilation

General Instructions: Complete space 6 if this work is a "changed version," "compilation," or "derivative work," and if it incorporates one or more earlier works that have already been published or registered for copyright, or that have fallen into the public domain. A "compilation" is defined as "a work formed by the collection and assembling of preexisting materials or of data that are selected, coordinated, or arranged in such a way that the resulting work as a whole constitutes an original work of authorship." A "derivative work" is "a work based on one or more preexisting works." Examples of derivative works include musical arrangements, dramatizations, translations, abridgments, condensations, motion picture versions.or "any other form in which a work may be recast, transformed, or adapted." Derivative works also include works "consisting of editorial revisions, annotations, or other modifications" if these changes, as a whole, represent an original work of authorship.

Preexisting Material (space 6a): Complete this space and space 6b for derivative works. In this space identify the preexisting work that has been recast, transformed, or adapted. For example, the preexisting material might be: "French version of Hugo's 'Le Roi s'amuse'." Do not complete this space for compilations.

Material Added to This Work (space 6b): Give a brief, general statement of the additional new material covered by the copyright claim for which registration is sought. In the case of a derivative work, identify this new material. Examples: "Arrangement for piano and orchestra"; "Dramatization for television"; "New film version"; "Revisions throughout; Act III completely new." If the work is a compilation, give a brief, general statement describing both the material that has been compiled and the compilation itself. Example: "Compilation of 19th Century Military Songs."

7,8,9 SPACE 7, 8, 9: Fee, Correspondence, Certification, Return Address

Deposit Account: If you maintain a Deposit Account in the Copyright Office, identify it in space 7. Otherwise leave the space blank and send the fee of $20 with your application and deposit.

Correspondence (space 7): This space should contain the name, address, area code, and telephone number of the person to be consulted if correspondence about this application becomes necessary.

Certification (space 8): The application cannot be accepted unless it bears the date and the handwritten signature of the author or other copyright claimant, or of the owner of exclusive right(s), or of the duly authorized agent of the author, claimant, or owner of exclusive right(s).

Address for Return of Certificate (space 9): The address box must be completed legibly since the certificate will be returned in a window envelope.

MORE INFORMATION

How To Register a Recorded Work: If the musical or dramatic work that you are registering has been recorded (as a tape, disk, or cassette), you may choose either copyright application Form PA or Form SR, Performing Arts or Sound Recordings, depending on the purpose of the registration.

Form PA should be used to register the underlying musical composition or dramatic work. Form SR has been developed specifically to register a "sound recording" as defined by the Copyright Act—a work resulting from the "fixation of a series of sounds," separate and distinct from the underlying musical or dramatic work. Form SR should be used when the copyright claim is limited to the sound recording itself. (In one instance, Form SR may also be used to file for a copyright registration for both kinds of works—see (4) below.) Therefore:

(1) File Form PA if you are seeking to register the musical or dramatic work, not the "sound recording," even though what you deposit for copyright purposes may be in the form of a phonorecord.

(2) File Form PA if you are seeking to register the audio portion of an audiovisual work, such as a motion picture soundtrack; these are considered integral parts of the audiovisual work.

(3) File Form SR if you are seeking to register the "sound recording" itself, that is, the work that results from the fixation of a series of musical, spoken, or other sounds, but not the underlying musical or dramatic work.

(4) File Form SR if you are the copyright claimant for both the underlying musical or dramatic work and the sound recording, and you prefer to register both on the same form.

(5) File both forms PA and SR if the copyright claimant for the underlying work and sound recording differ, or you prefer to have separate registration for them.

"Copies" and "Phonorecords": To register for copyright, you are required to deposit "copies" or "phonorecords." These are defined as follows:

Musical compositions may be embodied (fixed) in "copies," objects from which a work can be read or visually perceived, directly or with the aid of a machine or device, such as manuscripts, books, sheet music, film, and videotape. They may also be fixed in "phonorecords," objects embodying fixations of sounds, such as tapes and phonograph disks, commonly known as phonograph records. For example, a song (the work to be registered) can be reproduced in sheet music ("copies") or phonograph records ("phonorecords"), or both.

FORM PA
UNITED STATES COPYRIGHT OFFICE

REGISTRATION NUMBER

PA   PAU

EFFECTIVE DATE OF REGISTRATION

Month   Day   Year

DO NOT WRITE ABOVE THIS LINE. IF YOU NEED MORE SPACE, USE A SEPARATE CONTINUATION SHEET.

1

TITLE OF THIS WORK ▼

PREVIOUS OR ALTERNATIVE TITLES ▼

NATURE OF THIS WORK ▼ See instructions

2

a NAME OF AUTHOR ▼

DATES OF BIRTH AND DEATH
Year Born ▼   Year Died ▼

Was this contribution to the work a "work made for hire"?
☐ Yes
☐ No

AUTHOR'S NATIONALITY OR DOMICILE
Name of Country
OR { Citizen of ▶ _____
{ Domiciled in ▶ _____

WAS THIS AUTHOR'S CONTRIBUTION TO THE WORK
Anonymous? ☐ Yes ☐ No
Pseudonymous? ☐ Yes ☐ No
If the answer to either of these questions is "Yes," see detailed instructions.

NATURE OF AUTHORSHIP Briefly describe nature of the material created by this author in which copyright is claimed. ▼

NOTE

Under the law, the "author" of a "work made for hire" is generally the employer, not the employee (see instructions). For any part of this work that was "made for hire" check "Yes" in the space provided, give the employer (or other person for whom the work was prepared) as "Author" of that part, and leave the space for dates of birth and death blank.

b NAME OF AUTHOR ▼

DATES OF BIRTH AND DEATH
Year Born ▼   Year Died ▼

Was this contribution to the work a "work made for hire"?
☐ Yes
☐ No

AUTHOR'S NATIONALITY OR DOMICILE
Name of Country
OR { Citizen of ▶ _____
{ Domiciled in ▶ _____

WAS THIS AUTHOR'S CONTRIBUTION TO THE WORK
Anonymous? ☐ Yes ☐ No
Pseudonymous? ☐ Yes ☐ No
If the answer to either of these questions is "Yes," see detailed instructions.

NATURE OF AUTHORSHIP Briefly describe nature of the material created by this author in which copyright is claimed. ▼

c NAME OF AUTHOR ▼

DATES OF BIRTH AND DEATH
Year Born ▼   Year Died ▼

Was this contribution to the work a "work made for hire"?
☐ Yes
☐ No

AUTHOR'S NATIONALITY OR DOMICILE
Name of Country
OR { Citizen of ▶ _____
{ Domiciled in ▶ _____

WAS THIS AUTHOR'S CONTRIBUTION TO THE WORK
Anonymous? ☐ Yes ☐ No
Pseudonymous? ☐ Yes ☐ No
If the answer to either of these questions is "Yes," see detailed instructions.

NATURE OF AUTHORSHIP Briefly describe nature of the material created by this author in which copyright is claimed. ▼

3

a YEAR IN WHICH CREATION OF THIS WORK WAS COMPLETED This information must be given in all cases.
◀ Year

b DATE AND NATION OF FIRST PUBLICATION OF THIS PARTICULAR WORK Complete this information ONLY if this work has been published.
Month ▶ _____ Day ▶ _____ Year ▶ _____ ◀ Nation

4

COPYRIGHT CLAIMANT(S) Name and address must be given even if the claimant is the same as the author given in space 2.▼

APPLICATION RECEIVED

ONE DEPOSIT RECEIVED

TWO DEPOSITS RECEIVED

REMITTANCE NUMBER AND DATE

DO NOT WRITE HERE OFFICE USE ONLY

See instructions before completing this space.

TRANSFER If the claimant(s) named here in space 4 are different from the author(s) named in space 2, give a brief statement of how the claimant(s) obtained ownership of the copyright.▼

MORE ON BACK ▶
• Complete all applicable spaces (numbers 5-9) on the reverse side of this page.
• See detailed instructions.
• Sign the form at line 8.

DO NOT WRITE HERE

EXAMINED BY

CHECKED BY

☐ CORRESPONDENCE
　　Yes

FORM PA

FOR
COPYRIGHT
OFFICE
USE
ONLY

DO NOT WRITE ABOVE THIS LINE. IF YOU NEED MORE SPACE, USE A SEPARATE CONTINUATION SHEET.

PREVIOUS REGISTRATION Has registration for this work, or for an earlier version of this work, already been made in the Copyright Office?

☐ Yes ☐ No If your answer is "Yes," why is another registration being sought? (Check appropriate box) ▼

a. ☐ This is the first published edition of a work previously registered in unpublished form.

b. ☐ This is the first application submitted by this author as copyright claimant.

c. ☐ This is a changed version of the work, as shown by space 6 on this application.

If your answer is "Yes," give: Previous Registration Number ▼　　　　Year of Registration ▼

DERIVATIVE WORK OR COMPILATION Complete both space 6a & 6b for a derivative work; complete only 6b for a compilation.

a. Preexisting Material Identify any preexisting work or works that this work is based on or incorporates. ▼

b. Material Added to This Work Give a brief, general statement of the material that has been added to this work and in which copyright is claimed. ▼

See instructions
before completing
this space

DEPOSIT ACCOUNT If the registration fee is to be charged to a Deposit Account established in the Copyright Office, give name and number of Account.
Name ▼　　　　　　　　　　　　　　　　　　Account Number ▼

CORRESPONDENCE Give name and address to which correspondence about this application should be sent. Name/Address/Apt/City/State/Zip ▼

Be sure to
give your
daytime phone
◀ number

Area Code & Telephone number ▶

CERTIFICATION* I, the undersigned, hereby certify that I am the
Check only one ▼

☐ author

☐ other copyright claimant

☐ owner of exclusive right(s)

☐ authorized agent of _____
　　　　　　　Name of author or other copyright claimant, or owner of exclusive right(s) ▲

of the work identified in this application and that the statements made
by me in this application are correct to the best of my knowledge.

Typed or printed name and date ▼ If this application gives a date of publication in space 3, do not sign and submit it before that date.

date ▶

Handwritten signature (X) ▼

MAIL
CERTIFI-
CATE TO

Name ▼

Certificate
will be
mailed in
window
envelope

Number/Street/Apartment Number ▼

City/State/ZIP ▼

YOU MUST
• Complete all necessary spaces
• Sign your application in space 8
SEND ALL 3 ELEMENTS
IN THE SAME PACKAGE:
1. Application form
2. Nonrefundable $20 filing fee
in check or money order
payable to Register of Copyrights
3. Deposit material
MAIL TO
Register of Copyrights
Library of Congress
Washington, D.C. 20559

* 17 U S C § 506(e) Any person who knowingly makes a false representation of a material fact in the application for copyright registration provided for by section 409, or in any written statement filed in
connection with the application shall be fined not more than $2 500

⊘Filling Out Application Form TX

Detach and read these instructions before completing this form.
Make sure all applicable spaces have been filled in before you return this form.

———BASIC INFORMATION———

When to Use This Form: Use Form TX for registration of published or unpublished non-dramatic literary works, excluding periodicals or serial issues. This class includes a wide variety of works: fiction, nonfiction, poetry, textbooks, reference works, directories, catalogs, advertising copy, compilations of information, and computer programs. For periodicals and serials, use Form SE.

Deposit to Accompany Application: An application for copyright registration must be accompanied by a deposit consisting of copies or phonorecords representing the entire work for which registration is to be made. The following are the general deposit requirements as set forth in the statute:

Unpublished Work: Deposit one complete copy (or phonorecord).

Published Work: Deposit two complete copies (or one phonorecord) of the best edition.

Work First Published Outside the United States: Deposit one complete copy (or phonorecord) of the first foreign edition.

Contribution to a Collective Work: Deposit one complete copy (or phonorecord) of the best edition of the collective work.

The Copyright Notice: For works first published on or after March 1, 1989, the law provides that a copyright notice in a specified form "may be placed on all publicly distributed copies from which the work can be visually per-

ceived." Use of the copyright notice is the responsibility of the copyright owner and does not require advance permission from the Copyright Office. The required form of the notice for copies generally consists of three elements: (1) the symbol "©," or the word "Copyright," or the abbreviation "Copr."; (2) the year of first publication; and (3) the name of the owner of copyright. For example: "© 1993 Jane Cole." The notice is to be affixed to the copies "in such manner and location as to give reasonable notice of the claim of copyright." Works first published prior to March 1, 1989, **must** carry the notice or risk loss of copyright protection.

For information about notice requirements for works published before March 1, 1989, or other copyright information, write: Information Section, LM-401, Copyright Office, Library of Congress, Washington, D.C. 20559.

PRIVACY ACT ADVISORY STATEMENT Required by the Privacy Act of 1974 (Public Law 93-579)

AUTHORITY FOR REQUESTING THIS INFORMATION:
● Title 17, U.S.C., Secs. 409 and 410

FURNISHING THE REQUESTED INFORMATION IS:
● Voluntary

BUT IF THE INFORMATION IS NOT FURNISHED:
● It may be necessary to delay or refuse registration
● You may not be entitled to certain relief, remedies, and benefits provided in chapters 4 and 5 of title 17, U.S.C.

PRINCIPAL USES OF REQUESTED INFORMATION:
● Establishment and maintenance of a public record
● Examination of the application for compliance with legal requirements

OTHER ROUTINE USES:
● Public inspection and copying
● Preparation of public indexes
● Preparation of public catalogs of copyright registrations
● Preparation of search reports upon request

NOTE:
● No other advisory statement will be given you in connection with this application
● Please keep this statement and refer to it if we communicate with you regarding this application

———LINE-BY-LINE INSTRUCTIONS———
Please type or print using black ink.

1 SPACE 1: Title

Title of This Work: Every work submitted for copyright registration must be given a title to identify that particular work. If the copies or phonorecords of the work bear a title or an identifying phrase that could serve as a title, transcribe that wording *completely* and *exactly* on the application. Indexing of the registration and future identification of the work will depend on the information you give here.

Previous or Alternative Titles: Complete this space if there are any additional titles for the work under which someone searching for the registration might be likely to look or under which a document pertaining to the work might be recorded.

Publication as a Contribution: If the work being registered is a contribution to a periodical, serial, or collection, give the title of the contribution in the "Title of this Work" space. Then, in the line headed "Publication as a Contribution," give information about the collective work in which the contribution appeared.

2 SPACE 2: Author(s)

General Instructions: After reading these instructions, decide who are the "authors" of this work for copyright purposes. Then, unless the work is a "collective work," give the requested information about every "author" who contributed any appreciable amount of copyrightable matter to this version of the work. If you need further space, request Continuation sheets. In the case of a collective work such as an anthology, collection of essays, or encyclopedia, give information about the author of the collective work as a whole.

Name of Author: The fullest form of the author's name should be given. Unless the work was "made for hire," the individual who actually created the work is its "author." In the case of a work made for hire, the statute provides that "the employer or other person for whom the work was prepared is considered the author."

What is a "Work Made for Hire"? A "work made for hire" is defined as (1) "a work prepared by an employee within the scope of his or her employment"; or (2) "a work specially ordered or commissioned for use as a contribution to a collective work, as a part of a motion picture or other audiovisual work, as a translation, as a supplementary work, as a compilation, as an instructional text, as a test, as answer material for a test, or as an atlas, if the parties expressly agree in a written instrument signed by them that the works shall be considered a work made for hire." If you have checked "Yes" to indicate that the work was "made for hire," you must give the full legal name of the employer (or other person for whom the work was prepared). You may also include the name of the employee along with the name of the employer (for example: "Elster Publishing Co., employer for hire of John Ferguson").

"Anonymous" or "Pseudonymous" Work: An author's contribution to a work is "anonymous" if that author is not identified on the copies or phonorecords of the work. An author's contribution to a work is "pseudonymous" if that author is identified on the copies or phonorecords under a fictitious name. If the work is "anonymous" you may: (1) leave the line blank; or (2) state "anonymous" on the line; or (3) reveal the author's identity. If the work is "pseudonymous" you may: (1) leave the line blank; or (2) give the pseudonym and identify it as such (for example: "Huntley Haverstock, pseudonym"); or (3) reveal the author's name, making clear which is the real name and which is the pseudonym (for example, "Judith Barton, whose pseudonym is Madeline Elster"). However, the citizenship or domicile of the author must be given in all cases.

Dates of Birth and Death: If the author is dead, the statute requires that the year of death be included in the application unless the work is anonymous or pseudonymous. The author's birth date is optional but is useful as a form of identification. Leave this space blank if the author's contribution was a "work made for hire."

Author's Nationality or Domicile: Give the country of which the author is a citizen or the country in which the author is domiciled. Nationality or domicile **must** be given in all cases.

Nature of Authorship: After the words "Nature of Authorship," give a brief general statement of the nature of this particular author's contribution to the work. Examples: "Entire text"; "Coauthor of entire text"; "Chapters 11-14"; "Editorial revisions"; "Compilation and English translation"; "New text."

3 SPACE 3: Creation and Publication

General Instructions: Do not confuse "creation" with "publication." Every application for copyright registration must state "the year in which creation of the work was completed." Give the date and nation of first publication only if the work has been published.

Creation: Under the statute, a work is "created" when it is fixed in a copy or phonorecord for the first time. Where a work has been prepared over a period of time, the part of the work existing in fixed form on a particular date constitutes the created work on that date. The date you give here should be the year in which the author completed the particular version for which registration is now being sought, even if other versions exist or if further changes or additions are planned.

Publication: The statute defines "publication" as "the distribution of copies or phonorecords of a work to the public by sale or other transfer of ownership, or by rental, lease, or lending"; a work is also "published" if there has been an "offering to distribute copies or phonorecords to a group of persons for purposes of further distribution, public performance, or public display." Give the full date (month, day, year) when, and the country where, publication first occurred. If first publication took place simultaneously in the United States and other countries, it is sufficient to state "U.S.A."

4 SPACE 4: Claimant(s)

Name(s) and Address(es) of Copyright Claimant(s): Give the name(s) and address(es) of the copyright claimant(s) in this work even if the claimant is the same as the author. Copyright in a work belongs initially to the author of the work (including, in the case of a work made for hire, the employer or other person for whom the work was prepared). The copyright claimant is either the author of the work or a person or organization to whom the copyright initially belonging to the author has been transferred.

Transfer: The statute provides that, if the copyright claimant is not the author, the application for registration must contain "a brief statement of how the claimant obtained ownership of the copyright." If any copyright claimant named in space 4 is not an author named in space 2, give a brief statement explaining how the claimant(s) obtained ownership of the copyright. Examples: "By written contract"; "Transfer of all rights by author"; "Assignment"; "By will." Do not attach transfer documents or other attachments or riders.

5 SPACE 5: Previous Registration

General Instructions: The questions in space 5 are intended to show whether an earlier registration has been made for this work and, if so, whether there is any basis for a new registration. As a general rule, only one basic copyright registration can be made for the same version of a particular work.

Same Version: If this version is substantially the same as the work covered by a previous registration, a second registration is not generally possible unless: (1) the work has been registered in unpublished form and a second registration is now being sought to cover this first published edition; or (2) someone other than the author is identified as copyright claimant in the earlier registration, and the author is now seeking registration in his or her own name. If either of these two exceptions apply, check the appropriate box and give the earlier registration number and date. Otherwise, do not submit Form TX; instead, write the Copyright Office for information about supplementary registration or recordation of transfers of copyright ownership.

Changed Version: If the work has been changed and you are now seeking registration to cover the additions or revisions, check the last box in space 5, give the earlier registration number and date, and complete both parts of space 6 in accordance with the instructions below.

Previous Registration Number and Date: If more than one previous registration has been made for the work, give the number and date of the latest registration.

6 SPACE 6: Derivative Work or Compilation

General Instructions: Complete space 6 if this work is a "changed version," "compilation," or "derivative work" and if it incorporates one or more earlier works that have already been published or registered for copyright or that have fallen into the public domain. A "compilation" is defined as "a work formed by the collection and assembling of preexisting materials or of data that are selected, coordinated, or arranged in such a way that the resulting work as a whole constitutes an original work of authorship." A "derivative work" is "a work based on one or more preexisting works." Examples of derivative works include translations, fictionalizations, abridgments, condensations, or "any other form in which a work may be recast, transformed, or adapted." Derivative works also include works "consisting of editorial revisions, annotations, or other modifications" if these changes, as a whole, represent an original work of authorship.

Preexisting Material (space 6a): For derivative works, complete this space and space 6b. In space 6a identify the preexisting work that has been recast, transformed, or adapted. An example of preexisting material might be: "Russian version of Goncharov's 'Oblomov'." Do not complete space 6a for compilations.

Material Added to This Work (space 6b): Give a brief, general statement of the new material covered by the copyright claim for which registration is sought. **Derivative work** examples include: "Foreword, editing, critical annotations"; "Translation"; "Chapters 11-17." If the work is a **compilation**, describe both the compilation itself and the material that has been compiled. Example: "Compilation of certain 1917 Speeches by Woodrow Wilson." A work may be both a derivative work and compilation, in which case a sample statement might be: "Compilation and additional new material."

7 SPACE 7: Manufacturing Provisions

Due to the expiration of the Manufacturing Clause of the copyright law on June 30, 1986, this space has been deleted.

8 SPACE 8: Reproduction for Use of Blind or Physically Handicapped Individuals

General Instructions: One of the major programs of the Library of Congress is to provide Braille editions and special recordings of works for the exclusive use of the blind and physically handicapped. In an effort to simplify and speed up the copyright licensing procedures that are a necessary part of this program, section 710 of the copyright statute provides for the establishment of a voluntary licensing system to be tied in with copyright registration. Copyright Office regulations provide that you may grant a license for such reproduction and distribution solely for the use of persons who are certified by competent authority as unable to read normal printed material as a result of physical limitations. The license is entirely voluntary, nonexclusive, and may be terminated upon 90 days notice.

How to Grant the License: If you wish to grant it, check one of the three boxes in space 8. Your check in one of these boxes together with your signature in space 10 will mean that the Library of Congress can proceed to reproduce and distribute under the license without further paperwork. For further information, write for Circular 63.

9,10,11 SPACE 9,10,11: Fee, Correspondence, Certification, Return Address

Fee: The Copyright Office has the authority to adjust fees at 5-year intervals, based on changes in the Consumer Price Index. The next adjustment is due in 1996. Please contact the Copyright Office after July 1995 to determine the actual fee schedule.

Deposit Account: If you maintain a Deposit Account in the Copyright Office, identify it in space 9. Otherwise leave the space blank and send the fee of $20 with your application and deposit.

Correspondence (space 9) This space should contain the name, address, area code, and telephone number of the person to be consulted if correspondence about this application becomes necessary.

Certification (space 10): The application can not be accepted unless it bears the date and the **handwritten signature** of the author or other copyright claimant, or of the owner of exclusive right(s), or of the duly authorized agent of author, claimant, or owner of exclusive right(s).

Address for Return of Certificate (space 11): The address box must be completed legibly since the certificate will be returned in a window envelope.

FORM TX
For a Literary Work
UNITED STATES COPYRIGHT OFFICE

REGISTRATION NUMBER

TX	TXU

EFFECTIVE DATE OF REGISTRATION

Month	Day	Year

DO NOT WRITE ABOVE THIS LINE. IF YOU NEED MORE SPACE, USE A SEPARATE CONTINUATION SHEET.

1 TITLE OF THIS WORK ▼

PREVIOUS OR ALTERNATIVE TITLES ▼

PUBLICATION AS A CONTRIBUTION If this work was published as a contribution to a periodical, serial, or collection, give information about the collective work in which the contribution appeared.   **Title of Collective Work ▼**

If published in a periodical or serial give: Volume ▼   Number ▼   Issue Date ▼   On Pages ▼

2
a NAME OF AUTHOR ▼

DATES OF BIRTH AND DEATH
Year Born ▼   Year Died ▼

Was this contribution to the work a "work made for hire"?
☐ Yes
☐ No

AUTHOR'S NATIONALITY OR DOMICILE
Name of Country
OR { Citizen of ▶ _____
Domiciled in ▶ _____

WAS THIS AUTHOR'S CONTRIBUTION TO THE WORK
Anonymous?   ☐ Yes ☐ No
Pseudonymous? ☐ Yes ☐ No
If the answer to either of these questions is "Yes," see detailed instructions.

NATURE OF AUTHORSHIP Briefly describe nature of material created by this author in which copyright is claimed. ▼

NOTE

Under the law, the "author" of a "work made for hire" is generally the employer, not the employee (see instructions). For any part of this work that was "made for hire" check "Yes" in the space provided, give the employer (or other person for whom the work was prepared) as "Author" of that part, and leave the space for dates of birth and death blank.

b NAME OF AUTHOR ▼

DATES OF BIRTH AND DEATH
Year Born ▼   Year Died ▼

Was this contribution to the work a "work made for hire"?
☐ Yes
☐ No

AUTHOR'S NATIONALITY OR DOMICILE
Name of Country
OR { Citizen of ▶ _____
Domiciled in ▶ _____

WAS THIS AUTHOR'S CONTRIBUTION TO THE WORK
Anonymous?   ☐ Yes ☐ No
Pseudonymous? ☐ Yes ☐ No
If the answer to either of these questions is "Yes," see detailed instructions.

NATURE OF AUTHORSHIP Briefly describe nature of material created by this author in which copyright is claimed. ▼

NAME OF AUTHOR ▼

DATES OF BIRTH AND DEATH
Year Born ▼   Year Died ▼

Was this contribution to the work a "work made for hire"?
☐ Yes
☐ No

AUTHOR'S NATIONALITY OR DOMICILE
Name of Country
OR { Citizen of ▶ _____
Domiciled in ▶ _____

WAS THIS AUTHOR'S CONTRIBUTION TO THE WORK
Anonymous?   ☐ Yes ☐ No
Pseudonymous? ☐ Yes ☐ No
If the answer to either of these questions is "Yes," see detailed instructions.

NATURE OF AUTHORSHIP Briefly describe nature of material created by this author in which copyright is claimed. ▼

3
a YEAR IN WHICH CREATION OF THIS WORK WAS COMPLETED This information must be given in all cases. ◀ Year

b DATE AND NATION OF FIRST PUBLICATION OF THIS PARTICULAR WORK
Complete this information ONLY if this work has been published.
Month ▶ _____ Day ▶ _____ Year ▶ _____
◀ Nation

4 COPYRIGHT CLAIMANT(S) Name and address must be given even if the claimant is the same as the author given in space 2. ▼

See instructions before completing this space.

TRANSFER If the claimant(s) named here in space 4 is (are) different from the author(s) named in space 2, give a brief statement of how the claimant(s) obtained ownership of the copyright. ▼

DO NOT WRITE HERE / OFFICE USE ONLY
APPLICATION RECEIVED
ONE DEPOSIT RECEIVED
TWO DEPOSITS RECEIVED
FUNDS RECEIVED

MORE ON BACK ▶ • Complete all applicable spaces (numbers 5-11) on the reverse side of this page.
• See detailed instructions.   • Sign the form at line 10.

DO NOT WRITE HERE
Page 1 of _____ pages

EXAMINED BY	FORM TX
CHECKED BY	
☐ CORRESPONDENCE Yes	FOR COPYRIGHT OFFICE USE ONLY

DO NOT WRITE ABOVE THIS LINE. IF YOU NEED MORE SPACE, USE A SEPARATE CONTINUATION SHEET.

PREVIOUS REGISTRATION Has registration for this work, or for an earlier version of this work, already been made in the Copyright Office?
☐ Yes ☐ No If your answer is "Yes," why is another registration being sought? (Check appropriate box) ▼
a. ☐ This is the first published edition of a work previously registered in unpublished form.
b. ☐ This is the first application submitted by this author as copyright claimant.
c. ☐ This is a changed version of the work, as shown by space 6 on this application.
If your answer is "Yes," give: **Previous Registration Number ▼**   **Year of Registration ▼**

5

DERIVATIVE WORK OR COMPILATION Complete both space 6a and 6b for a derivative work; complete only 6b for a compilation.
a. Preexisting Material Identify any preexisting work or works that this work is based on or incorporates. ▼

b. Material Added to This Work Give a brief, general statement of the material that has been added to this work and in which copyright is claimed. ▼

See instructions
before completing
this space.

6

—space deleted—

7

REPRODUCTION FOR USE OF BLIND OR PHYSICALLY HANDICAPPED INDIVIDUALS A signature on this form at space 10 and a check in one
of the boxes here in space 8 constitutes a non-exclusive grant of permission to the Library of Congress to reproduce and distribute solely for the blind and physically
handicapped and under the conditions and limitations prescribed by the regulations of the Copyright Office: (1) copies of the work identified in space 1 of this
application in Braille (or similar tactile symbols); or (2) phonorecords embodying a fixation of a reading of that work; or (3) both.

a ☐ Copies and Phonorecords   b ☐ Copies Only   c ☐ Phonorecords Only

See instructions.

8

DEPOSIT ACCOUNT If the registration fee is to be charged to a Deposit Account established in the Copyright Office, give name and number of Account.
Name ▼   **Account Number ▼**

9

CORRESPONDENCE Give name and address to which correspondence about this application should be sent. Name/Address/Apt/City/State/ZIP ▼

Be sure to
give your
daytime phone
◀ number

Area Code and Telephone Number ▶

CERTIFICATION* I, the undersigned, hereby certify that I am the
Check only one ▶
☐ author
☐ other copyright claimant
☐ owner of exclusive right(s)
☐ authorized agent of _____
of the work identified in this application and that the statements made
by me in this application are correct to the best of my knowledge.
Name of author or other copyright claimant, or owner of exclusive right(s) ▲

Typed or printed name and date ▼ If this application gives a date of publication in space 3, do not sign and submit it before that date.
_____ date ▶ _____

Handwritten signature (X) ▼

10

MAIL CERTIFI- CATE TO	Name ▼	**YOU MUST:** • Complete all necessary spaces • Sign your application in space 10
Certificate will be mailed in window envelope	Number/Street/Apartment Number ▼ City/State/ZIP ▼	**SEND ALL 3 ELEMENTS IN THE SAME PACKAGE:** 1. Application form 2. Nonrefundable $20 filing fee in check or money order payable to *Register of Copyrights* 3. Deposit material **MAIL TO:** Register of Copyrights Library of Congress Washington, D.C. 20559-6000

The Copyright Office
has the authority to ad-
just fees at 5-year intervals, based on changes
in the Consumer Price
Index. The next adjust-
ment is due in 1996.
Please contact the
Copyright Office after
July 1995 to determine
the actual fee schedule.

11

*17 U.S.C. § 506(e): Any person who knowingly makes a false representation of a material fact in the application for copyright registration provided for by section 409, or in any written statement filed in connection
with the application, shall be fined not more than $2,500.

July 1993—300,000   ♻ PRINTED ON RECYCLED PAPER   ☆U.S. GOVERNMENT PRINTING OFFICE: 1993-342-582/80,019

⊘Filling Out Application Form CA

Detach and read these instructions before completing this form.
Make sure all applicable spaces have 'been filled in before you return this form.

BASIC INFORMATION

Use Form CA When:

An earlier registration has been completed in the Copyright Office; and

Some of the facts given in that registration are incorrect or incomplete; and

You want to place the correct or complete facts on record.

Purpose of Supplementary Copyright Registration:
As a rule, only one basic copyright registration can be made for the same work. To take care of cases where information in the basic registration turns out to be incorrect or incomplete, section 408(d) of the copyright law provides for "the filing of an application for supplementary registration, to correct an error in a copyright registration or to amplify the information given in a registration."

Who May File:
Once basic registration has been made for a work, any author or other copyright claimant or owner of any exclusive right in the work or the duly authorized agent of any such author, other claimant, or owner who wishes to correct or amplify the information given in the basic registration may submit Form CA.

Please Note:

Do not use Form CA to correct errors in statements on the copies or phonorecords of the work in question or to reflect changes in the content of the work. If the work has been changed substantially, you should consider making an entirely new registration for the revised version to cover the additions or revisions.

Do not use Form CA as a substitute for renewal registration. Renewal of copyright cannot be accomplished by using Form CA. For information on renewal of copyright, write the Copyright Office for Circular 15.

Do not use Form CA as a substitute for recording a transfer of copyright or other document pertaining to rights under a copyright. Recording a document under section 205 of the statute gives all persons constructive notice of the facts stated in the document and may have other important consequences in cases of infringement or conflicting transfers. Supplementary registration does not have that legal effect.

For information on recording a document, request Circular 12 from the Copyright Office. To record a document in the Copyright Office, request the Document Cover Sheet.

How to Apply for Supplementary Registration:

First: Study the information on this page to make sure that filing an application on Form CA is the best procedure to follow in your case.

Second: Read the back of this page for the specific instructions on filling out Form CA. Before starting to complete the form, make sure that you have all of the necessary detailed information from the certificate of the basic registration.

Third: Complete all applicable spaces on this form following the line-by-line instructions on the back of this page. Use a typewriter or print the information in black ink.

Fourth: Detach this sheet and send your completed Form CA to: Register of Copyrights, Library of Congress, Washington, D.C. 20559-6000. Unless you have a Deposit Account in the Copyright Office, your application must be accompanied by a nonrefundable filing fee in the form of a check or money order for $20 payable to: *Register of Copyrights*. Do not send copies, phonorecords, or supporting documents with your application. They cannot be made part of the record of a supplementary registration.

What Happens When a Supplementary Registration is Made?
When a supplementary registration is completed, the Copyright Office will assign it a new registration number in the appropriate registration category and will issue a certificate of supplementary registration under that number. The basic registration will not be cancelled. The two registrations will stand in the Copyright Office records. The supplementary registration will have the effect of calling the public's attention to a possible error or omission in the basic registration and of placing the correct facts or the additional information on official record.

LINE-BY-LINE INSTRUCTIONS

Please type or print using black ink.

A PART A: Identification of Basic Registration

General Instructions: The information in this part identifies the basic registration that will be corrected or amplified. Even if the purpose of filing Form CA is to change one of these items, each item must agree exactly with the information as it already appears in the basic registration, that is, as it appears in the registration you wish to correct. Do not give any new information in this part.

Title of Work: Give the title as it appears in the basic registration.

Registration Number: Give the registration number (the series of numbers preceded by one or more letters) that appears in the upper right-hand corner of the certificate of registration.

Registration Date: Give the year when the basic registration was completed.

Name(s) of Author(s) and Name(s) of Copyright Claimant(s): Give all of the names as they appear in the basic registration.

B PART B: Correction

General Instructions: Complete this part **only** if information in the basic registration **was incorrect at the time that basic registration was made.** Leave this part blank and complete Part C, instead, if your purpose is to add, update, or clarify information rather than to rectify an actual error.

Location and Nature of Incorrect Information: Give the line number and the heading or description of the space in the basic registration where the error occurs. Example: "Line number 3 . . . Citizenship of author."

Incorrect Information as it Appears in Basic Registration: Transcribe the incorrect statement exactly as it appears in the basic registration, even if you have already given this information in Part A.

Corrected Information: Give the statement as it should have appeared in the application of the basic registration.

Explanation of Correction: You may need to add an explanation to clarify this correction.

C PART C: Amplification

General Instructions: Complete this part if you want to provide any of the following: (1) information that was omitted at the time of basic registration; (2) changes in facts other than ownership but including changes such as title or address of claimant, that have occurred since the basic registration; or (3) explanations clarifying information in the basic registration.

Location and Nature of Information to be Amplified: Give the line number and the heading or description of the space in the basic registration where the information to be amplified appears.

Amplified Information: Give a statement of the additional, updated, or explanatory information as clearly and succinctly as possible.

Explanation of Amplification: You should add an explanation of the amplification if it is necessary to clarify the amplification.

D,E,F,G PARTS D,E,F,G: Continuation, Fee, Certification, Return Address

Continuation (Part D): Use this space if you do not have enough room in Parts B or C.

Fee: The Copyright Office has the authority to adjust fees at 5-year intervals, based on changes in the Consumer Price Index. The next adjustment is due in 1996. Please contact the Copyright Office after July 1995 to determine the actual fee schedule.

Deposit Account and Mailing Instructions (Part E): If you maintain a Deposit Account in the Copyright Office, identify it in Part E. Otherwise, you will need to send the nonrefundable filing fee of $20 with your form. The space headed "Correspondence" should contain the name, address, and telephone number with area code of the person to be consulted if correspondence about the form becomes necessary.

Certification (Part F): The application is not acceptable unless it bears the handwritten signature of the author, or other copyright claimant, or of the owner of exclusive right(s), or of the duly authorized agent of such author, claimant, or owner.

Address for Return of Certificate (Part G): The address box must be completed legibly, since the certificate will be returned in a window envelope.

FORM CA

For Supplementary Registration
UNITED STATES COPYRIGHT OFFICE

REGISTRATION NUMBER

| TX | TXU | PA | PAU | VA | VAU | SR | SRU | RE |

EFFECTIVE DATE OF SUPPLEMENTARY REGISTRATION

Month   Day   Year

DO NOT WRITE ABOVE THIS LINE. IF YOU NEED MORE SPACE, USE A SEPARATE CONTINUATION SHEET.

A

TITLE OF WORK ▼

REGISTRATION NUMBER OF THE BASIC REGISTRATION ▼   YEAR OF BASIC REGISTRATION ▼

NAME(S) OF AUTHOR(S) ▼   NAME(S) OF COPYRIGHT CLAIMANT(S) ▼

B

LOCATION AND NATURE OF INCORRECT INFORMATION IN BASIC REGISTRATION ▼

Line Number . . . . Line Heading or Description . . . .

INCORRECT INFORMATION AS IT APPEARS IN BASIC REGISTRATION ▼

CORRECTED INFORMATION ▼

EXPLANATION OF CORRECTION ▼

C

LOCATION AND NATURE OF INFORMATION IN BASIC REGISTRATION TO BE AMPLIFIED ▼

Line Number . . . . Line Heading or Description . . . .

AMPLIFIED INFORMATION ▼

EXPLANATION OF AMPLIFIED INFORMATION ▼

MORE ON BACK ▶ • Complete all applicable spaces (D -G) on the reverse side of this page.
• See detailed instructions.   • Sign the form at space F.

DO NOT WRITE HERE
Page 1 of _____ pages

FORM CA RECEIVED

FORM CA

FUNDS RECEIVED DATE

EXAMINED BY

CHECKED BY

CORRESPONDENCE ☐

REFERENCE TO THIS REGISTRATION ADDED TO
BASIC REGISTRATION ☐ YES ☐ NO

FOR
COPYRIGHT
OFFICE
USE
ONLY

DO NOT WRITE ABOVE THIS LINE. IF YOU NEED MORE SPACE, USE A SEPARATE CONTINUATION SHEET.

CONTINUATION OF: (Check which) ☐ PART B OR ☐ PART C

D

DEPOSIT ACCOUNT: If the registration fee is to be charged to a Deposit Account established in the Copyright Office, give name and number of Account.

Name _____

Account Number _____

CORRESPONDENCE: Give name and address to which correspondence about this application should be sent.

Name _____

Address _____ (Apt)

(City) (State) (ZIP)

Area Code and Telephone Number ▶ _____

Be sure to give your daytime phone number

E

CERTIFICATION* I, the undersigned, hereby certify that I am the: (Check one)
☐ author ☐ other copyright claimant ☐ owner of exclusive right(s) ☐ duly authorized agent of _____
(Name of author or other copyright claimant, or owner of exclusive right(s) ▲

of the work identified in this application and that the statements made by me in this application are correct to the best of my knowledge.

Typed or printed name ▼

Date ▼

Handwritten signature (X) ▼

F

MAIL
CERTIFI-
CATE TO

Name ▼

Number/Street/Apt ▼

Certificate
will be
mailed in
window
envelope

City/State/ZIP ▼

YOU MUST:
• Complete all necessary spaces
• Sign your application in space F
SEND ALL ELEMENTS IN THE SAME PACKAGE:
1. Application form
2. Nonrefundable $20 filing fee in check or money order payable to *Register of Copyrights*
MAIL TO:
Register of Copyrights
Library of Congress
Washington, D.C. 20559-6000

The Copyright Office has the authority to adjust fees at 5-year intervals, based on changes in the Consumer Price Index. The next adjustment is due in 1996. Please contact the Copyright Office after July 1995 to determine the actual fee schedule.

G

*17 U.S.C. § 506(e): Any person who knowingly makes a false representation of a material fact in the application for copyright registration provided for by section 409, or in any written statement filed in connection with the application, shall be fined not more than $2,500.

December 1993—25,000

☆U.S. GOVERNMENT PRINTING OFFICE: 1993-301-241/80,049

✒️Filling Out Application Form RE

Detach and read these instructions before completing this form.
Make sure all applicable spaces have been filled in before you return this form.

————————BASIC INFORMATION————————

How to Register a Renewal Claim:

First: Study the information on this page and make sure you know the answers to two questions:
(1) What is the renewal filing period in your case?
(2) Who can claim the renewal?

Second: Read through the specific instructions for filling out Form RE. Before starting to complete the form, make sure that the copyright is now eligible for renewal, that you are authorized to file a renewal claim, and that you have all of the information about the copyright you will need.

Third: Complete all applicable spaces on Form RE, following the line-by-line instructions. Use typewriter or print the information in black ink.

Fourth: Detach this sheet and send your completed Form RE to: Register of Copyrights, Library of Congress, Washington, D.C. 20559. Unless you have a Deposit Account in the Copyright Office, your application must be accompanied by a check or money order for $20, payable to: *Register of Copyrights*. Do not send copies, phonorecords, or supporting documents with your renewal application unless specifically requested to do so by the Copyright Office.

What Is Renewal of Copyright?

For works copyrighted before January 1, 1978, the copyright law provides a first term of copyright protection lasting 28 years. These works were required to be renewed within strict time limits in order to obtain a second term of copyright protection lasting 47 years. If copyright originally secured before January 1, 1964, was not renewed at the proper time, copyright protection expired permanently at the end of the 28th year and could not be renewed.
Public Law 102-307, enacted June 26, 1992, amended the copyright law to extend automatically the term of copyrights secured between January 1, 1964, and December 31, 1977, to a further term of 47 years. This recent legislation makes renewal registration optional. The first term of copyright protection expires on December 31st of the 28th year of the original term of the copyright and the 47-year renewal term automatically vests in the party entitled to claim renewal as of that date.

Some Basic Points About Renewal:

(1) A work is eligible for renewal registration at the beginning of the 28th year of the first term of copyright.
(2) There is no requirement to make a renewal filing in order to extend the original 28-year copyright term to the full term of 75 years; however, there are some benefits from making a renewal registration during the 28th year of the original term. (For more information, write to the Copyright Office for Circular 15.)
(3) Only certain persons who fall into specific categories named in the law can claim renewal.
(4) For works originally copyrighted on or after January 1, 1978, the copyright law has eliminated all renewal requirements and established a single copyright term and different methods for computing the duration of a copyright. (For further information, write the Copyright Office for Circular 15a.)

Renewal Filing Period:

The amended copyright statute provides that, in order to register a renewal copyright, the renewal application and fee must be received in the Copyright Office
—within the last (28th) calendar year before the expiration of the original term of copyright or
—at any time during the renewed and extended term of 47 years.

To determine the filing period for renewal in your case:
(1) First, find out the date of original copyright for the work. (In the case of works originally registered in unpublished form, the date of copyright is the date of registration; for published works, copyright begins on the date of first publication.)
(2) Then add 28 years to the year the work was originally copyrighted.
Your answer will be the calendar year during which the copyright will become eligible for renewal. Example: A work originally copyrighted on April 19, 1966, will be eligible for renewal in the calendar year 1994.

To renew a copyright during the original copyright term, the renewal application and fee **must** be received in the Copyright Office within 1 year prior to the expiration of the original copyright. All terms of the original copyright run through the end of the 28th calendar year making the period for renewal registration during the original term from December 31st of the 27th year of the copyright through December 31st of the following year.

Who May Claim Renewal:

Renewal copyright may be claimed only by those persons specified in the law. Except in the case of four specific types of works, the law gives the right to claim renewal to the individual author of the work, regardless of who owned the copyright during the original term. If the author is dead, the statute gives the right to claim renewal to certain of the author's beneficiaries (widow and children, executors, or next of kin, depending on the circumstances). The present owner (proprietor) of the copyright is entitled to claim renewal only in four specified cases as explained in more detail on the reverse of this page.

PRIVACY ACT ADVISORY STATEMENT Required by the Privacy Act of 1974 (Public Law 93-579)	PRINCIPAL USES OF REQUESTED INFORMATION: ● Establishment and maintenance of a public record ● Examination of the application for compliance with legal requirements
AUTHORITY FOR REQUESTING THIS INFORMATION: ● Title 17, U.S.C., Sec. 304	OTHER ROUTINE USES: ● Public inspection and copying ● Preparation of public indexes
FURNISHING THE REQUESTED INFORMATION IS: ● Voluntary	● Preparation of public catalogs of copyright registrations ● Preparation of search reports upon request
BUT IF THE INFORMATION IS NOT FURNISHED: ● It may be necessary to delay or refuse renewal registration ● If renewal registration is not made before expiration of the original copyright term, ownership of the renewal term may be affected	NOTE: ● No other advisory statement will be given you in connection with this application ● Please keep this statement and refer to it if we communicate with you regarding this application

————LINE-BY-LINE INSTRUCTIONS————
Please type or print using black ink.

1 SPACE 1: Renewal Claimant(s)

General Instructions: In order for this application to result in a valid renewal, space 1 must identify one or more of the persons who are entitled to renew the copyright under the statute. Give the full name and address of each claimant, with a statement of the basis of each claim, using the wording given in these instructions.

For registration in the 28th year of the original copyright term, the renewal claimant is the individual(s) or entity who is entitled to claim renewal copyright on the date filed.

For registration after the 28th year of the original copyright term, the renewal claimant is the individual(s) or entity who is entitled to claim renewal copyright on December 31st of the 28th year.

Persons Entitled to Renew:
A. The following persons may claim renewal in all types of works except those enumerated in Paragraph B below:

1. The author, if living. State the claim as: *the author*

2. The widow, widower, and/or children of the author, if the author is not living. State the claim as:
the widow (widower) of the author ....
(Name of author)
and/or the child (children) of the deceased author ....
(Name of author)

3. The author's executor(s), if the author left a will and if there is no surviving widow, widower, or child. State the claim as:
the executor(s) of the author ....
(Name of author)

4. The next of kin of the author, if the author left no will and if there is no surviving widow, widower, or child. State the claim as:
the next of kin of the deceased author ... . *there being no will.*
(Name of author)

B. In the case of the following four types of works, the proprietor (owner of the copyright at the time of renewal registration) may claim renewal:

1. Posthumous work (a work published after the author's death as to which no copyright assignment or other contract for exploitation has occurred during the author's lifetime). State the claim as: *proprietor of copyright in a posthumous work.*

2. Periodical, cyclopedic, or other composite work. State the claim as: *proprietor of copyright in a composite work.*

3. "Work copyrighted by a corporate body otherwise than as assignee or licensee of the individual author." State the claim as: *proprietor of copyright in a work copyrighted by a corporate body otherwise than as assignee or licensee of the individual author.* (This type of claim is considered appropriate in relatively few cases.)

4. Work copyrighted by an employer for whom such work was made for hire. State the claim as: *proprietor of copyright in a work made for hire.*

2 SPACE 2: Work Renewed

General Instructions: This space is to identify the particular work being renewed. The information given here should agree with that appearing in the certificate of original registration.

Title: Give the full title of the work, together with any subtitles or descriptive wording included with the title in the original registration. In the case of a musical composition, give the specific instrumentation of the work.

Renewable Matter: Copyright in a new version of a previously published or copyrighted work (such as an arrangement, translation, dramatization, compilation, or work republished with new matter) covers only the additions, changes, or other new material appearing for the first time in that version. If this work was a new version, state in general the new matter upon which copyright was claimed.

Contribution to Periodical, Serial, or other Composite Work: Separate renewal registration is possible for a work published as a contribution to a periodical, serial, or other composite work, whether the contribution was copyrighted independently or as part of the larger work in which it appeared. Each contribution published in a separate issue ordinarily requires a separate renewal registration. However, the law provides an alternative, permitting groups of periodical contributions by the same individual author to be combined under a single renewal application and fee in certain cases.

If this renewal application covers a single contribution, give all of the requested information in space 2. If you are seeking to renew a group of contributions, include a reference such as "See space 5" in space 2 and give the requested information about all of the contributions in space 5.

3 SPACE 3: Author(s)

General Instructions: The copyright secured in a new version of a work is independent of any copyright protection in material published earlier. The only "authors" of a new version are those who contributed copyrightable matter to it. Thus, for renewal purposes, the person who wrote the original version on which the new work is based cannot be regarded as an "author" of the new version, unless that person also contributed to the new matter.

Authors of Renewable Matter: Give the full names of all authors who contributed copyrightable matter to this particular version of the work.

4 SPACE 4: Facts of Original Registration

General Instructions: Each item in space 4 should agree with the information appearing in the original registration for the work. If the work being renewed is a single contribution to a periodical or composite work that was not separately registered, give information about the particular issue in which the contribution appeared. You may leave this space blank if you are completing space 5.

Original Registration Number: Give the full registration number, which is a series of numerical digits, preceded by one or more letters. The registration number appears in the upper right hand corner of the front of the certificate of registration.

Original Copyright Claimant: Give the name in which ownership of the copyright was claimed in the original registration.

Date of Publication or Registration: Give only one date. If the original registration gave a publication date, it should be transcribed here; otherwise the registration was for an unpublished work, and the date of registration should be given.
NOTE: An original registration is not required but there are supplemental deposit requirements. You may call or write the Renewals Section for details. Phone 202-707-8180, or FAX 202-707-3849.
Renewals Section, LM 449
Copyright Office
Library of Congress
Washington, D.C. 20559

5 SPACE 5: Group Renewals

General Instructions: A renewal registration using a single application and $20 fee can be made for a group of works if all of the following statutory conditions are met: (1) all of the works were written by the same author, who is named in space 3 and who is or was an individual (not an employer for hire); (2) all of the works were first published as contributions to periodicals (including newspapers) and were copyrighted on their first publication; (3) the renewal claimant or claimants and the basis of claim or claims, as stated in space 1, are the same for all of the works; (4) the renewal application and fee are received not less than 27 years after the 31st day of December of the calendar year in which all of the works were first published; and (5) the renewal application identifies each work separately, including the periodical containing it and the date of first publication.

Time Limits for Group Renewals: To be renewed as a group, all of the contributions must have been first published during the same calendar year. For example, suppose six contributions by the same author were published on April 1, 1965, July 1, 1965, November 1, 1965, February 1, 1966, July 1, 1966, and March 1, 1967. The three 1965 copyrights can be combined and renewed at any time during 1993, and the two 1966 copyrights can be renewed as a group during 1994, but the 1967 copyright must be renewed by itself, in 1995.

Identification of Each Work: Give all of the requested information for each contribution. The registration number should be that for the contribution itself if it was separately registered, and the registration number for the periodical issue if it was not.

6,7,8 SPACE 6,7,8: Fee, Correspondence, Certification, Return Address

Fee: The Copyright Office has the authority to adjust fees at 5-year intervals, based on changes in the Consumer Price Index. The next adjustment is due in 1996. Please contact the Copyright Office after July 1995 to determine the actual fee schedule.

Deposit Account and Correspondence (Space 6): If you maintain a Deposit Account in the Copyright Office, identify it in space 6. Otherwise, you will need to send the renewal registration fee of $20 with your form. The space headed "Correspondence" should contain the name and address of the person to be consulted if correspondence about the form becomes necessary.

Certification (Space 7): The renewal application is not acceptable unless it bears the handwritten signature of the renewal claimant or the duly authorized agent of the renewal claimant.

Address for Return of Certificate (Space 8): The address box must be completed legibly, since the certificate will be returned in a window envelope.

FORM RE
For Renewal of a Work
UNITED STATES COPYRIGHT OFFICE

REGISTRATION NUMBER

EFFECTIVE DATE OF RENEWAL REGISTRATION

| Month | Day | Year |

DO NOT WRITE ABOVE THIS LINE. IF YOU NEED MORE SPACE, USE A SEPARATE CONTINUATION SHEET(RE/CON).

1 RENEWAL CLAIMANT(S), ADDRESS(ES), AND STATEMENT OF CLAIM ▼ (See Instructions)

1
Name ....
Address ....
Claiming as ....
(Use appropriate statement from instructions)

2
Name ....
Address ....
Claiming as ....

3
Name ....
Address ....
Claiming as ....

2 TITLE OF WORK IN WHICH RENEWAL IS CLAIMED ▼

RENEWABLE MATTER ▼

PUBLICATION AS A CONTRIBUTION If this work was published as a contribution to a periodical, serial, or other composite work, give information about the collective work in which the contribution appeared.   **Title of Collective Work ▼**

If published in a periodical or serial give: **Volume ▼**   **Number ▼**   **Issue Date ▼**

3 AUTHOR(S) OF RENEWABLE MATTER ▼

4 ORIGINAL REGISTRATION NUMBER ▼ ORIGINAL COPYRIGHT CLAIMANT ▼

ORIGINAL DATE OF COPYRIGHT
If the original registration for this work was made in published form, give:
DATE OF PUBLICATION: _____
(Month)   (Day)   (Year)
OR
If the original registration for this work was made in unpublished form, give:
DATE OF REGISTRATION: _____
(Month)   (Day)   (Year)

MORE ON BACK ▶   • Complete all applicable spaces (numbers 5-8) on the reverse side of this page.
• See detailed instructions.   • Sign the form at space 7.

DO NOT WRITE HERE
Page 1 of .... pages

RENEWAL APPLICATION RECEIVED	FORM RE
CORRESPONDENCE ☐ YES	FOR COPYRIGHT OFFICE USE ONLY
EXAMINED BY	
CHECKED BY	

DO NOT WRITE ABOVE THIS LINE. IF YOU NEED MORE SPACE, USE A SEPARATE CONTINUATION SHEET (RE/CON).

RENEWAL FOR GROUP OF WORKS BY SAME AUTHOR: To make a single registration for a group of works by the same individual author published as contributions to periodicals (see instructions), give full information about each contribution. If more space is needed, request continuation sheet (Form RE/CON).

5

1
Title of Contribution: ....
Title of Periodical: .... Vol: .... No: .... Issue Date: ....
Date of Publication: .... Registration Number: ....
(Month) (Day) (Year)

2
Title of Contribution: ....
Title of Periodical: .... Vol: .... No: .... Issue Date: ....
Date of Publication: .... Registration Number: ....
(Month) (Day) (Year)

3
Title of Contribution: ....
Title of Periodical: .... Vol: .... No: .... Issue Date: ....
Date of Publication: .... Registration Number: ....
(Month) (Day) (Year)

4
Title of Contribution: ....
Title of Periodical: .... Vol: .... No: .... Issue Date: ....
Date of Publication: .... Registration Number: ....
(Month) (Day) (Year)

DEPOSIT ACCOUNT: If the registration fee is to be charged to a Deposit Account established in the Copyright Office, give name and number of Account.

Name _____

_____

Account Number _____

CORRESPONDENCE: Give name and address to which correspondence about this application should be sent.

Name _____

Address _____ (Apt)

(City) (State) (ZIP)

Area Code and Telephone Number ▶ _____

Be sure to give your daytime phone ◀ number

6

CERTIFICATION* I, the undersigned, hereby certify that I am the: (Check one)
☐ renewal claimant ☐ duly authorized agent of _____
(Name of renewal claimant) ▲
of the work identified in this application and that the statements made by me in this application are correct to the best of my knowledge.

Typed or printed name ▼

Date ▼

☞ Handwritten signature (X) ▼

7

MAIL CERTIFI-CATE TO

Certificate will be mailed in window envelope

Name ▼

Number/Street/Apt ▼

City/State/ZIP ▼

YOU MUST:
• Complete all necessary spaces
• Sign your application in space 7

SEND ALL ELEMENTS IN THE SAME PACKAGE:
1. Application form
2. Nonrefundable $20 filing fee in check or money order payable to *Register of Copyrights*

MAIL TO:
Register of Copyrights
Library of Congress
Washington, D.C. 20559

The Copyright Office has the authority to adjust fees at 5-year intervals, based on changes in the Consumer Price Index. The next adjustment is due in 1996. Please contact the Copyright Office after July 1995 to determine the actual fee schedule.

8

*17 U.S.C. § 506(e): Any person who knowingly makes a false representation of a material fact in the application for copyright registration provided for by section 409, or in any written statement filed in connection with the application, shall be fined not more than $2,500.

April 1993—40,000

☆U.S. GOVERNMENT PRINTING OFFICE: 1993-342-581/60.513

Copyright Appendix III

International Copyright Protection

There is no such thing as an international copyright that will automatically protect an author's writings throughout the entire world. Protection against unauthorized use in a particular country depends, basically, on the national laws of that country. However, most countries do offer protection to foreign works under certain *conditions,* and these conditions have been greatly simplified by international copyright treaties and conventions. For a list of countries which maintain copyright relations with the United States, request Circular 38a.

The United States belongs to both of the global, multilateral copyright treaties—the Univeral Copyright Convention (UCC) and the Bern Convention for the Protection of Literary and Artistic Works. The United States was a founding member of the UCC, which came into force on September 16, 1955. Generally, both a work by a national or domiciliary of a country that is a member of the UCC, and a work first published in a UCC country, may claim protection under the UCC. If the work bears the notice of copyright in the form and position specified by the UCC, this notice will satisfy and substitute for any other formalities a UCC member country would otherwise impose as a condition of copyright. A UCC notice should consist of the symbol © accompanied by the name of the copyright proprietor and the year of first publication of the work.

By joining the Bern Convention on March 1, 1989, the United States gained protection for its authors in all member nations of the Bern Union with which the United States formerly had either no copyright relations, or bilateral treaty arrangements. Members of the Bern Union agree to a certain *minimum* level of copyright protection, and also to treat nationals of other member countries like their *own* nationals, for purposes of copyright. A work first published in the United States or another Bern Union country (or first published in a non-Bern country, followed by publication within thirty days in a Bern Union country) is eligible for protection in *all* Bern member countries. There are no special requirements. For information on the legislation implementing the Bern Convention, request Circular 93 from the Copyright Office.

An author who wishes protection for his or her work in a particular country should first find out the extent of protection of foreign works in that country. If possible, this should be done before the work is published *anywhere,* since protection may often depend on the facts existing at the time and in the place of first publication.

If the country in which protection is sought is a party to one of the international copyright conventions, the work may generally be protected by complying with the conditions of that convention. Even if the work *cannot* be brought under an international convention, protection under the specific provisions of the country's *national laws* may still be possible. Some countries, unfortunately, offer little or no copyright protection for foreign works.

United States Copyright and Foreign Origin of the Work

Copyright protection is available from the U.S. Copyright Office for all *unpublished* works, regardless of the nationality or domicile of the author. In the case of works created in foreign lands, however, certain conditions apply. *Published* works are eligible for copyright protection in the United States if *any one* of the following conditions is met:

- On the date of the first publication of the work, one or more of its authors is either (1) a national or domiciliary of the United States; (2) a national, domiciliary, or sovereign authority of a foreign nation that is a party to a copyright treaty to which the United States is also a party; or (3) a stateless person.
- The work is first published in the United States or in a foreign nation that on the date of first publication is a party to the Universal Copyright Convention.

- The work comes within the scope of a Presidential proclamation.
- The work was first published on or after March 1, 1989, in a foreign nation that on the date of first publication was a party to the Bern Convention.
- The work is not first published in a country party to the Bern Convention, but is then published within thirty days of the first publication in a country party to the Bern Convention (on or after March 1, 1989).
- The work, first published on or after March 1, 1989, is a pictorial, graphic, or sculptural work that is incorporated in a permanent structure located in the United States.
- The work, first published on or after March 1, 1989, is a published audiovisual work, and all the authors are legal entities with headquarters in the United States.

Library of Congress Catalog Numbers

A Library of Congress Catalog Card Number is different from a copyright registration number. The Cataloging in Publication (CIP) Division of the Library of Congress is responsible for assigning LC Catalog Card Numbers and is operationally separate from the Copyright Office. A book may be registered in or deposited with the Copyright Office but not necessarily cataloged and added to the Library's collections. For information about obtaining an LC Catalog Card Number, contact the CIP Division, Library of Congress, Washington, DC 20540. For information on International Standard Book Numbering (ISBN), write to ISBN Agency, R. R. Bowker Co., 245 West 17th St., New York, NY 10011. For information on International Standard Serial Numbering (ISSN), write to Library of Congress, National Serials Data Program, Washington, DC 20540.

Use of Mandatory Deposit to Satisfy Registration Requirements

For works published in the United States, the Copyright Act contains a provision under which a single deposit can be made to satisfy both the deposit requirements for the Library, and the registration requirements. In order to have this dual effect, the copies or phonorecords must be accompanied by the prescribed application and filing fee.

PART **IV**

Trademarks

CHAPTER 10

Trademark Basics

In its most general sense, a *trademark* is a word, a symbol, a design, a combination of word and design, a slogan, or even a distinctive sound that identifies and distinguishes the goods or services of one party from those of another. In other words, it is a *mark* used in *trade* by a particular business or individual. If the mark is used to identify a service, rather than a trade, it is called a *service* mark. Normally, a trademark appears on the product or its packaging, whereas a service mark is used in advertising, to identify the owner's services. Since the concepts and rules behind both are identical, the term trademark is used to refer to *both* trademarks *and* service marks.

Until now, we've been discussing patents and copyrights. As we've seen, a copyright gives protection for an artistic or literary work, and a patent gives protection for an invention. The trademark is fundamentally different from both.

One of the major differences is in the duration of the protection afforded. A *patent* lasts seventeen years, a *copyright* for the author's life plus 50 years. A *trademark,* however, lasts as long as the mark continues to perform a source-identifying function: theoretically, forever.

Trademark Rights

The protection afforded by a trademark is similar to that offered by copyrights and patents: A trademark owner has the exclusive right to use the device to identify particular products, *and* the right to prevent others from using a confusingly similar mark to identify the same or similar products.

When Would I Need a Trademark?

Obviously, if you intend to do business under a particular business name, you already *have* a trademark—in a sense. Anytime you produce or sell a particular product that you want your consumers to identify readily, you use a *brand name* (which is just another term for a trademark). Beyond the obvious, though, there are two situations under which it might be a good idea to design, register, and use a trademark for your work *as a supplement to* your existing intellectual property rights. The first is if you have a patented invention, since the trademark can provide further protection. The second is if you're seeking additional protection for a fictional character you regularly use in copyrighted work—since characters themselves *cannot* be copyrighted.

Trademarking a Product Name for a Patent

Why would a trademark be useful for your patented product? Throughout the section on patents, we used the example of a new type of paper clip. Now let's suppose you had a thriving business for your own paper clips. After seventeen years, you're still in business—but your patent has expired, and suddenly the market is flooded with imitations of your product. With no particular way for consumers to identify *your* paper clip, your business could fall dramatically.

Now, let's suppose that, along with the patent, you filed for and received a trademark for a distinctive logo for "Joe's Snappy Paper Clip." For the next seventeen years, as your business grew, your consumers would come to identify the product with your *trademark* name. Then, when the patent expired and the market was flooded with imitations, your trademark would still be in effect, and name recognition alone would help your business to continue thriving.

Some of you may be surprised that names like Kleenex and Jeep can be trademarks (both of these *are*). These trademarks have come into such general use that people now confuse the brand names with the product. This situation isn't necessarily desirable either, since consumers may pick up *any* brand of paper handkerchief and assume they're getting a "Kleenex."

Trademarking a Character Name

The second instance where trademarking might come in handy to supplement your other intellectual property rights is via a *character name*—but the protection afforded here is less clear. Generally, characters are considered copyrightable if they are "distinctively delineated," but while this protection was upheld for such famous characters as Tarzan and Hopalong Cassidy, the same argument did *not* work for the famous detective created by Dashiell Hammett, Sam Spade, when in 1954 a court ruled that, though Warner Bros. held the exclusive copyright on *a story containing* the character, they did not hold the exclusive copyright on *the character*—which could *not* be copyrighted.

Since then, in at least some cases, the courts have ruled that character names *can* be trademarked, and benefit from the protection thus afforded. The names (and images) of Mork and Mindy, Batman, and SPIDER-MAN all are court-upheld trademarks. Trademarking your character names will make it much easier to prevent companies from using them to sell unauthorized products.

The Source of Trademark Protection

Trademark rights arise from either of the following:

• The use of the mark
• A bona fide intention to use the mark

As in the case of copyright, federal trademark registration is *not* required in order for a trademark to be protected, and a trademark may be used *without* obtaining a registration. We strongly recommend, however, that you proceed with the registration process, since it greatly amplifies your protection in case of infringement.

Benefits of Registration

While federal registration isn't necessary for trademark protection, registration on what's called the Principal Register provides the following "Top Ten" advantages:

1. Nationwide priority as of the filing date of the application, which provides the registrant with a date of first use of the mark in commerce
2. The right to sue in federal court for trademark infringement
3. Recovery of profits, damages, and costs in a federal court infringement action, and the possibility of treble damages and attorneys' fees
4. Constructive notice of a claim of ownership—which eliminates a good-faith defense for a second party adopting the trademark subsequent to the registrant's date of registration
5. The right to deposit the registration with Customs in order to stop the importation of goods bearing an infringing mark
6. *Prima facie* evidence of the validity of the registration, the ownership of the mark, and the exclusive right to use the mark in commerce in connection with the goods or services specified in the certificate
7. The possibility of incontestability, in which case the registration constitutes conclusive evidence of the registrant's exclusive right, with certain limited exceptions, to use the registered mark in commerce
8. Limited grounds for attacking a registration once it is five years old
9. Availability of criminal penalties and treble damages in an action for counterfeiting a registered trademark
10. A basis for filing trademark applications in foreign countries

The Registration Process

An application for federal *registration* of a trademark is filed with the Patent and Trademark Office. As in the case of copyrights, the form is fairly straightforward, so it *is not* necessary to seek the assistance of a lawyer. (However, doing so *is* advisable. There are some sticky areas involved, so some legal help may be necessary.)

Before a trademark owner can file an application, however, he or she must accomplish *either* of the following:

- Use the mark on goods which are shipped or sold, or services which are rendered, in commerce regulated by Congress. Interstate and international commerce all are regulated by Congress, so a trademark must be used *at least once* in an interstate or international trade in order to achieve registration.
- Have a bona fide intention to use the mark in such commerce in relation to specific goods or services.

When an application is filed, it is reviewed to determine whether it meets the requirements for receiving a filing date. If the filing requirements are *not* met, the entire mailing, including the fee, is returned to the applicant. If the application *does* meet the filing requirements, it is assigned a serial number, and the applicant is sent a filing receipt.

Initial Determination

The first part of the registration process is a determination by a trademark examining attorney as to whether the mark may be registered. An initial determination (listing any statutory grounds for refusal, as well as any procedural informalities in the application) is issued about three months after filing. The applicant must respond to any objections to the application within six months, or the application will be considered abandoned. If, after reviewing the applicant's response, the Examining Attorney makes a final refusal of registration, the applicant may appeal to the Trademark Trial and Appeal Board, an administrative tribunal within the Patent and Trademark Office.

Approval

If the Examining Attorney approves the mark, the mark will be published in the *Trademark Official Gazette*, a weekly publication of the Patent and Trademark Office. Any other party then has thirty days either to oppose the registration or to request an extension of time to oppose it. An *opposition* is similar to proceedings in federal district courts, but is held before the

Trademark Trial and Appeal Board. If no opposition is filed, the application enters the next stage of the registration process.

If the mark published is based upon its actual use in commerce, a registration will be issued approximately twelve weeks after the date on which the mark was published.

Intent To Use

If the mark was published based upon the applicant's statement of a bona fide intention to use it in commerce, a *notice of allowance* will be issued about twelve weeks after the date on which the mark was published. The applicant then has six months from the date of the notice of allowance either (1) to use the mark in commerce and submit a *statement of use,* or (2) to request a six-month *extension of time* to file a statement of use. The applicant may request additional extensions of time *only* as noted in the instructions on the *back* of the form!

CHAPTER 11

Trademark Applications Up Close

A Trademark application consists of the following parts, or elements:

1. A written application form
2. A drawing of the mark
3. The required filing fee
4. Three specimens showing actual use of the mark on or in connection with the goods or services

The fourth element is necessary only if the application is filed based upon prior use of the mark in commerce. A separate application must be filed for each mark for which registration is requested.

Now let's take a closer look at each of the elements that make up a complete application. The written application for trademark registration can be found in Trademark Appendix I of this section. It is titled the Trademark/Service Mark Application, Principal Register, With Declaration. The back page of the form is printed upside down so that it may be affixed to the application file at the top and still be easily read.

The Written Application Form

The same application form may be used for either a trademark or service mark application. Additional forms may be photocopied. All applications must be written in English.

The following explanations cover the various headings that you will encounter.

The Heading

Identify (a) the mark and (b) the class number(s) of the goods or services for which registration is sought. The "class number" is part of the Patent and Trademark Office's administrative process. To find the appropriate number, refer to The International Schedule of Classes of Goods and Services, Trademark Appendix III of this section. If you can't find an appropriate class number on the list, this section may be left blank.

Applicant

The application must be filed in the name of the owner of the mark. Specify, if an individual, your name and citizenship; if a partnership, specify the names and citizenship of the general partners and the domicile of the partnership; if a corporation or association, specify the name under which it is incorporated and the state or foreign nation under the laws of which it is organized. Also indicate the applicant's post office address.

Identification of Goods or Services

State briefly the specific goods or services for which the mark is intended or already used. Use clear and precise language. For example: "women's clothing—namely, blouses and skirts" or "computer programs for use by accountants" or "retail food store services." Describe the goods the applicant sells, or the services the applicant renders, but not the medium in which the mark appears, which often is advertisements. "Advertising" in this context would identify a service rendered by advertising agencies. For example, a restaurateur would identify his service as "restaurant services," and not "menus, signs, etc.," which is the medium through which the mark is communicated.

Basis for Application

The applicant must check at least one of four boxes to specify the basis for filing the application. Usually an application is based on either (1) prior use of the mark in commerce (the first box) or (2) a bona fide intention to use the mark in commerce (the second box)—but not both. If both the first *and* second boxes are checked, the Patent and Trademark Office will not accept the application and will return it to the applicant without processing.

The last two boxes pertain to applications filed in the United States pursuant to international agreements and based upon applications or registrations in foreign countries. These are rarely used. For further information about foreign-based applications, the applicant may call the trademark information number listed in this book, or contact a private attorney.

If the applicant is using the mark in commerce in relation to all the goods or services listed in the application, check the first box, and then do each of the following:

- State the date on which the trademark was first used anywhere in the United States on the goods, or in connection with the services, specified in the application.
- State the date on which the trademark was first used on the specified goods, or in connection with the specified services, sold or shipped (or rendered) in a type of commerce which may be regulated by Congress.
- Describe the type of commerce in which the goods were sold or shipped, or services were rendered (for example: "interstate commerce" or "commerce between the United States and [specify foreign country]").
- Describe how the mark is used on the goods, or in connection with the services (for example: "The mark is used on labels which are affixed to the goods" or "The mark is used in advertisements for the services").

If the applicant has a bona fide intention to use the mark in commerce in relation to the goods or services specified in the application, check the second box. This would include situations where the mark has not been used at all, or where the mark has been used on the specified books or services only within a single state (intrastate commerce).

Execution

The application must be dated and signed on the back of the form. The Patent and Trademark Office will not accept an unsigned application and will return it to the applicant without processing. By signing the form, the applicant is swearing that all the information in the application is believed to be true. If the applicant is an individual, the individual must execute it; if joint applicants, all must execute it; if a partnership, one general partner must execute the application; and if a corporation or association, one officer of the organization must execute the application.

The Drawing

The second element of the complete trademark registration application is the drawing. It is a representation of the mark as actually used, or intended to be used, on the goods or services.

There are two types of drawings:

1. Typed drawings, which consist of words or phrases and can be created with the use of a typewriter
2. Special form drawings, which require more-complex designs that should be rendered by a competent artist

In either case, all drawings must be made upon pure white, durable, non-shiny paper that is 8½″ wide and 11″ long. One of the shorter sides of the sheet should be regarded as its top. The margins must be at least one inch from the sides and the bottom of the paper. There must also be at least one inch between the drawing of the mark and the heading.

Keep in mind that the drawing is different from the specimen. The specimens must be the actual tags or labels (for goods) or advertisements (for services) that evidence the use of the mark in commerce. The drawing, on the other hand, is a black-and-white, or typed, rendition of the mark. The drawing will be used in printing the mark both in the Official Gazette and on the registration certificate. A copy of the drawing is also filed in the paper records of the Trademark Search Library, to provide notice of the pending application.

Drawing Heading

Across the top of the drawing, beginning one inch from the top edge and not exceeding one-third of the sheet, the following must be listed, on separate lines:

- The applicant's name
- The applicant's post office address
- The goods or services specified in the application (or *typical* items thereof if there are *many* listed)
- In an application based on use in *commerce,* the date of first use of the mark anywhere in the U.S., *and* the date of first use of the mark in commerce
- On an application based on a *foreign* application, the filing date of the *foreign* application.

Typed Drawing

If the mark is only words, or words and numerals, and the applicant *does not* want the registration to be issued for a particular depiction of the words and/or numerals, the mark may simply be typed in capital letters in the center of the page.

Special-Form Drawing

This type of drawing must be used if the applicant *does* want the registration for the mark to be issued in a particular style, *or* if the mark contains a design element. The drawing of the mark must be done in *black ink,* either with an india-ink pen or by some process which will afford similarly satisfactory reproduction characteristics. *Every* line and letter, including words, must be black. Halftones and gray are *not* acceptable—and this applies to *all* lines, including any used for shading. All lines must be clean, sharp, and solid—not a one of them fine or crowded.

A photolithographic reproduction, printer's proof, or camera-ready copy may be used in place of the original. Photographs, however, are *not* acceptable. Photocopies are acceptable *only* if they produce an unusually clear and sharp black-and-white rendering. The use of white pigment or white tape to cover lines is *not* acceptable.

Size of Special Drawings

The preferred size of the drawing is 2½″ × 2½″, and in no case may it be larger than 4″ × 4″. The Patent and Trademark Office will *not* accept an application with a special-form drawing depicted larger than 4″ × 4″, and will return the application without processing. If the amount of detail in the mark precludes clear reduction to the required 4″ × 4″ size, such detail should not be *shown* in the drawing, but should be verbally *described* in the body of the application.

Use of Color in Special Drawings

Where color is a feature of a mark, the color or colors may be designated in the drawing by the linings shown in the accompanying chart on page 212.

Specimens (Examples of Use)

The third element of the application is the *specimen,* an example of actual use of the trademark in trade. (The specimen is *not* necessary for *intent-to-use* applicants.) For purposes of the specimen, and in general, trademarks may be placed:

- On the goods
- On the container for the goods
- On displays associated with the goods
- On tags or labels attached to the goods
- On documents associated with the goods or their sale (if the nature of the goods makes other placement impractical)

Likewise, service marks may appear in advertisements for the services, or in brochures about the services, or on business cards or stationery used in connection with the services.

For an application based on actual use of the mark in commerce, the applicant must furnish three such examples of use as outlined above, when the application is filed. The Patent and Trademark Office will *not* accept an application based on use in commerce without at least one specimen and will return it to the applicant without processing.

Nature of the Specimens

The three specimens may either be identical or examples of three different types of uses. For goods, the specimens should be actual labels, tags, containers, displays, etc. For services, specimens should be actual advertisements, brochures, store signs, or stationery (if the nature of the services is clear from the letterhead or body of the letter), and so on.

Specimens may not be larger than 8½" × 11" and must be capable of being arranged flat. Three-dimensional or bulky material is not acceptable. Photographs or other reproductions clearly and legibly showing the mark on the goods, or on displays associated with the goods, may be submitted *if* the manner of affixing the mark to the goods, *or* the nature of the goods, is such that specimens as described above *cannot* be submitted.

The Filing Fee

The fourth and final element of the complete application is the *filing fee*. This fee, which became effective on April 17, 1989, is $245.00 for each class of goods or services for which the application is made. (For a list of classes, refer to The International Schedule of Classes of Goods and Services, Trademark Appendix III of this section.) *At least* $200 must be submitted for the application to be given a filing date.

Payment should be made in United States *specie,* or coin in the general sense, treasury notes, national bank notes, post office money orders, or certified checks. Personal or business checks may also be submitted. Letters containing cash should be registered. Remittances made from foreign countries must be payable and immediately negotiable in the United States for the full amount of the fee required. The application fees are *nonrefundable.*

Further Requirements for Intent-to-Use Applicants

Applicants who allege only a bona fide *intention* to use a mark in commerce must make use of the mark in commerce *before* the registration will be issued. Once use of the mark begins, the applicant must submit:

- Specimens evidencing use
- A fee of $100 per class of goods or services in the application
- Either an Amendment to Allege Use, *or* a Statement of Use

The difference between the standard filing and the intent-to-use filing is entirely in the timing. Copies of each of these forms appear in Trademark Appendix I of this section. For assistance in filling out these forms, refer to the instructions and information on the back of each.

Extensions

Since the intent-to-use applicant is under a deadline to file a statement of use, the Patent and Trademark Office allows for an extension—under certain circumstances. The necessary form, entitled "Request for Extension of Time Under 37 CFR 2.89 to File a Statement of Use, With Declaration" can be obtained from a Patent and Trademark Office. For assistance in filling out the form, refer as usual to the instructions and other information on the back.

Grounds for Refusal of the Application

The examining attorney will *refuse*, for *any* of the following reasons, registration of the mark or term applied for:

1. If the mark does not function as a trademark to identify the goods or services as coming from a particular source—if, for example, it is merely ornamentation
2. If the mark is immoral, deceptive, or scandalous
3. If the mark may disparage or falsely suggest a connection with persons, institutions, beliefs, or national symbols, or bring them into contempt or disrepute
4. If the mark consists of or simulates the flag or coat of arms or other insignia of the United States, or of a State or municipality, or of any foreign nation
5. If the mark is the name, portrait, or signature of a particular living individual, unless he or she has given written consent; or is the name, signature, or portrait of a deceased President of the United States during the life of his widow, unless she has given her consent
6. If the mark so resembles a mark already registered as to be likely, when used on or in connection with the goods of the applicant, to cause confusion, or to cause mistake, or to deceive
7. If the mark is merely descriptive or deceptively misdescriptive of the goods or services

8. If the mark is primarily geographically descriptive or deceptively misdescriptive of the goods or services of the applicant
9. If the mark is primarily merely a surname

A mark will *not* be refused registration on the grounds listed in numbers 7, 8, and 9 *if* the applicant can show that, through use of the mark in commerce, the mark has become distinctive—so that it now identifies to the public the applicant's goods or services.

The Supplemental Register

It is possible that certain marks refused registration on the grounds listed above as numbers 1, 7, 8, and 9 may still be shown in what's called the *Supplemental Register.* This information source contains terms or designs considered capable of distinguishing the owner's goods or services, but that do not yet do so. A term or design cannot be considered for inclusion in this register unless it is in use in commerce in relation to the goods or services identified in the application, and an acceptable allegation of use has been submitted.

If a mark *is* featured in this register, the registrant may bring suit for trademark infringement in the federal courts, and use the registration as a basis for filing in some foreign countries. (None of the other benefits of federal registration listed earlier applies.) An applicant may file an application in the Principal Register and, if appropriate, amend the application to the Supplemental Register for no additional fee.

Trademark Search Library

The Patent and Trademark Office maintains a record of all active registrations and pending applications in what's called a *search library.* These records are open to the public so that they may be used to help determine whether a previously registered mark exists which could prevent the registration of an applicant's mark.

The Patent and Trademark Office cannot advise you of the availability of a particular mark *before* you file your application. If you are unable to visit the library and search the records yourself, you can, however, hire a private search company or law firm to do the work for you. Naturally, the Office cannot recommend any such companies. You can consult your telephone directory under Patent Trademark Searches or contact local bar associations for a list of attorneys specializing in trademark law.

The search library is located just outside Washington, D.C., at Crystal Plaza 2, 2nd Floor, 2011 Jefferson Davis Highway, Arlington, VA 22022. It is open to the public (free of charge) Monday through Friday, 8:00 A.M. to 5:30 P.M. during normal work weeks.

Who May File an Application

The owners of marks may file and prosecute their own applications for registration, or be represented by an attorney. The Patent and Trademark Office cannot help select an attorney.

Notice

Once a federal registration is issued, the registrant may give notice of registration by using:

- The ® symbol
- Either the phrase "Registered in U.S. Patent and Trademark Office" *or* the phrase "Reg. U.S. Pat. & Tm. Off."

Although registration symbols may not be lawfully used prior to registration, many trademark owners use a TM or SM (if the mark identifies a service) symbol to indicate a claim of ownership, even if no federal trademark application is pending.

Renewal and Maintenance Terms

Once you have your registration, to maintain your protection effectively you will be required to renew it every ten years. You should also note that between the fifth and sixth years following the date of registration, you must file an affidavit stating that the mark is currently in use in commerce. If no affidavit is filed, the registration will be *canceled*.

Conclusion

Trademarks can provide an effective form of protection not only for general brand and trade names but also, in a supplemental sense, for patented inventions and copyrighted characters. The ease of the application process and the cost of the filing fees, unlike those of the patent process, places trademark registration protection in the hands of practically anyone. (In cases of infringement, the same general rules apply for trademarks as do for copyrights and patents.) In any event, if infringement occurs, or someone accuses *you* of infringement, the best course of action is to be aware of your rights and responsibilities, and then to consult a knowledgeable attorney.

Now that we've reviewed the basics of the four major areas of protection, trade secrets, patents, and copyrights and trademarks, it is important that you understand how these areas work together to form a protective web for your intellectual property. In each area, ultimately, there are as many exceptions as there are rules, as many gray areas as there are clear-cut standards, and as many cases lost as won.

Keep in mind that the intellectual property laws are meant to be flexible to an extent, in order to provide greater protection and fairness on a situation-to-situation basis. For example, while you might think that the formula for Coca-Cola falls under patent and/or copyright protection, *neither* is the case. It is protected first as a trade secret, but also as a trademark.

Most of the time you won't have problems with someone stealing your ideas. Nine times out of ten, the purpose of contracts and laws is not to protect us when things are going *well*, but to safeguard us when things go *badly*. That said, we hope you will use the information here to help you make your ideas a lucrative reality.

Trademark Application Forms

FILING REQUIREMENTS

WARNING: BEFORE COMPLETING AN APPLICATION, READ THE INSTRUCTIONS CAREFULLY AND STUDY THE EXAMPLES PROVIDED. ERRORS OR OMISSIONS MAY RESULT IN THE DENIAL OF A FILING DATE AND THE RETURN OF APPLICATION PAPERS, *OR* THE DENIAL OF REGISTRATION AND FORFEITURE OF THE FILING FEE.

To receive a filing date, the applicant *must* provide all of the following:
1.   a **written application form**;
2.   a **drawing** of the mark on a separate piece of paper;
3.   the required filing fee ($200.00 per class); and
4.   if the application is filed based upon prior use of the mark in commerce, **three specimens** for each class of goods or services. The specimens must show actual use of the mark with the goods or services.

1.   WRITTEN APPLICATION FORM [PTO FORM 1478]

The application must be in English. A separate application must be filed for each mark the applicant wishes to register. PTO Form 1478 included in the back of this booklet may be used for either a trademark or service mark application. It may be photocopied for your convenience. See the examples of completed applications on pages 12 and 13 with references to the following line-by-line instructions.

LINE-BY-LINE INSTRUCTIONS FOR FILLING OUT PTO FORM 1478, ENTITLED "TRADEMARK/SERVICE MARK APPLICATION, PRINCIPAL REGISTER, WITH DECLARATION"

Space 1 -- The Mark

Indicate the mark (e.g., "THEORYTEC" or "PINSTRIPES AND DESIGN"). This should agree with the mark shown on the drawing page.

Space 2 -- Classification

It is *not* necessary to fill in this box. The PTO will determine the proper International Classification based upon the identification of the goods and services in the application. However, if the applicant knows the International Class number(s) for the goods and services, the applicant may place the number(s) in this box. The International Classes are listed inside of the back cover.

Space 3 -- The Owner of the Mark

The name of the owner of the mark must be entered in this box. The application must be filed in the name of the owner of the mark or the application will be void, and the applicant will forfeit the filing fee. Thus, it is very important to determine who owns the mark before applying. The owner of the mark is the party who controls the nature and quality of the goods sold, or services rendered, under the mark. The owner may be an individual, a partnership, a corporation, or an association or similar firm. If the applicant is a corporation, the applicant's name is the name under which it is incorporated.

Space 4 -- The Owner's Address

Enter the applicant's business address. If the applicant is an individual, enter either the applicant's business or home address.

Space 5 -- Entity Type and Citizenship/Domicile

The applicant must check the box which indicates the type of entity applying. In addition, in the blank following the box, the applicant must specify the following information:

Space 5(a) -- for an **individual**, the applicant's national citizenship;

Space 5(b) -- for a **partnership**, the names and national citizenship of the general partners and the state where the partnership is organized (if a U.S. partnership) or country (if a foreign partnership);

Space 5(c) -- for a **corporation**, the state of incorporation (if a U.S. corporation), or country (if a foreign corporation); or

Space 5(d) -- for another type of entity, specify the nature of the entity and the state where it is organized (if in the U.S.) or country where it is organized (if a foreign entity).

Space 6 -- Identification of the Goods and/or Services

In this blank the applicant must state the specific goods and services for which registration is sought and with which the applicant has actually used the mark in commerce, or in the case of an "intent-to-use" application, has a bona fide intention to use the mark in commerce. Use clear and concise terms specifying the actual goods and services by their common commercial names. Use language that would be readily understandable to the general public. For example, if the applicant uses or intends to use the mark to identify "candy," "word processors," "baseballs and baseball bats," "travel magazines," "dry cleaning services," "restaurant services" or "insurance agency services," the identification should clearly and concisely list each such item. If the applicant uses indefinite terms, such as "accessories," "components," "devices," "equipment," "food," "materials," "parts," "systems," "products," or the like, then those words must be followed by the word "namely" and the goods or services listed by their common commercial name(s).

The applicant must be very careful when identifying the goods and services. Because the filing of an application establishes certain presumptions of rights as of the filing date, the application may *not* be amended later to add any products or services not within the scope of the identification. For example, the identification of "clothing" could be amended to "shirts and jackets," which narrows the scope, but could not be amended to "retail clothing store services," which would change the scope. Similarly, "physical therapy services" could not be changed to "medical services" because this would broaden the scope of the identification.

Moreover, the identification of goods and services must *not* describe the mode of use of the mark, such as on labels, stationery, menus, signs, containers or in advertising. There is another place on the application, called the "method-of-use clause," for this kind of information. (See information under Space 7a, fourth blank, described on the next page.) For example, in the identification of goods and services, the term "advertising" usually is intended to identify a *service* rendered by advertising agencies. Moreover, "labels," "menus," "signs" and "containers" are specific *goods*. If the applicant identifies these goods or services by mistake, the applicant may *not* amend the identification to the actual goods or services of the applicant. Thus, if the identification indicates "menus," it could not be amended to "restaurant services." Similarly, if the goods are identified as "containers or labels for jam," the identification could not be amended to "jam."

NOTE: If nothing appears in this blank, or if the identification does not identify any recognizable goods or services, the application will be denied a filing date and returned to the applicant. For example, if the applicant specifies the mark itself or wording such as "company name," "corporate name," or "company logo," and nothing else, the application will be denied a filing date and returned to the applicant. If the applicant identifies the goods and services too broadly as, for example, "advertising and business," "miscellaneous," "miscellaneous goods and services," or just "products," or "services," the application will also be denied a filing date and returned to the applicant.

Space 7 -- Basis for Filing

The applicant must check at least one of the four boxes to specify a basis for filing the application. The applicant should also fill in all blanks which follow the checked box(es). Usually an application is based upon either (1) prior use of the mark in commerce (the first box), or (2) a bona fide intention to use the mark in commerce (the second box). **You may *not* check both the first and second box. If both the first and second boxes are checked, the PTO will *not* accept the application and will return it to the applicant without processing.**

Space 7(a)

If the applicant is using the mark in commerce in relation to all of the goods and services listed in the application, check this first box and fill in the blanks.

In the **first blank** specify the date the trademark was first used with the goods and services in a type of commerce which may be regulated by Congress.

In the **second blank** specify the type of commerce, specifically a type of commerce which may be regulated by Congress, in which the goods were sold or shipped, or the services were rendered. For example, indicate "interstate commerce" (commerce between two or more states) or commerce between the United States and a specific foreign country, for example, "commerce between the U.S. and Canada."

In the **third blank** specify the date that the mark was first used anywhere in the U.S. with the goods or services specified in the application. This date will be the same as the date of first use in commerce unless the applicant made some use, for example, within a single state, before its first use in commerce.

In the **fourth blank** specify how the mark is placed on the goods or used with the services. This is referred to as the "method-of-use clause," and should not be confused with the identification of the goods and services described under Space 6. For example, in relation to goods, state "the mark is used on labels affixed to the goods," or "the mark is used on containers for the goods," whichever is accurate. In relation to services, state "the mark is used in advertisements for the services."

Space 7(b)

If the applicant has a bona fide intention to use the mark in commerce in relation to the goods or services specified in the application, check this second box and fill in the blank. The applicant should check this box if the mark has not been used at all or if the mark has been used on the specified goods or services only within a single state (intrastate commerce).

In the blank, state how the mark is intended to be placed on the goods or used with the services. For example, for goods, state "the mark will be used on labels affixed to the goods," or "the mark will be used on containers for the goods," whichever is accurate. For services, state "the mark will be used in advertisements for the services."

Spaces 7(c) and (d)

These spaces are usually used only by applicants in foreign countries who are filing in the United States under international agreements. These applications are less common. For further information about treaty-based applications, call the trademark information number listed in this booklet or contact a private attorney.

Space 8 -- Verification and Signature

The applicant must verify the truth and accuracy of the information in the application and must sign the application. The declaration in Space 8, on the back of the form, is for this purpose. If the application is not signed, the application will not be granted a filing date and will be returned to the applicant. If the application is not signed by an appropriate party, the application will be found void and the filing fee will be forfeited. Therefore, it is important that the proper person sign the application.

Who should sign?

- If the applicant is an individual, that individual must sign.

- If the applicant is a partnership, a general partner must sign.

- If the applicant is a corporation, association or similar organization, an officer of the corporation, association or organization must sign. An officer is a person who holds an office established in the articles of incorporation or the bylaws. Officers may not delegate this authority to nonofficers.

- If the applicants are joint applicants, all joint applicants must sign.

In addition to signing the application, the person who signs the application must indicate the date signed, provide a telephone number to be used if it is necessary to contact the applicant, and clearly print or type their name and position.

2. THE DRAWING PAGE

All applications must include a drawing page or the application will be denied a filing date and returned to the applicant. The PTO uses the drawing to file the mark in the PTO search records and to print the mark in the *Official Gazette* and on the registration.

The drawing must be on pure white, durable, non-shiny paper which is approximately 8½ (21.6 cm) inches wide by 11 (27.9 cm) inches long. There must be at least a one-inch margin on the sides, top and bottom of the page, and at least one inch between the heading and the display of the mark.

At the top of the drawing there must be a heading, listing on separate lines, the applicant's complete name, address, the goods and services specified in the application, and in applications based on use in commerce, the date of first use of the mark and the date of first use of the mark in commerce. This heading should be typewritten. If the drawing is in special form, the heading should include a description of the essential elements of the mark.

The drawing of the mark should appear at the center of the page. The drawing of the mark may be typewritten, as shown on page 15, or it may be in special form, as shown on page 14.

If the drawing is in **typewritten form**, the mark *must* be typed entirely in CAPITAL LETTERS. Capital letters must be used even if the mark, as used, includes lower-case letters.

If the applicant wishes to register a word mark in the form in which it is actually used or intended to be used in commerce, or any mark including a design, the applicant must submit a **special-form** drawing. In a special--form drawing, the mark *must not* be larger than 4 inches by 4 inches (10.3 cm by 10.3 cm). If the display of the mark is larger than 4 inches by 4 inches, the application will be denied a filing date and returned to the applicant. The mark in a special-form drawing *must be identical to the display of the mark on the specimens*. However, the drawing must appear only in black and white, with every line and letter black and clear. No color or gray is allowed. If the applicant wishes to indicate color, the applicant must use the color linings shown below. Most drawings do not indicate specific colors.

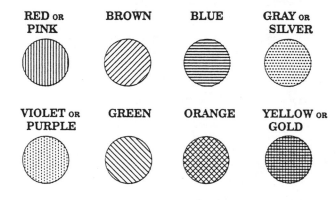

Be careful in preparing the drawing. While the applicant sometimes can make some minor changes to the mark after filing, the rules prohibit any material change to the mark in the drawing after filing.

3. FEES

Filing Fee

The application filing fee is $200.00 for each class of goods or services listed, for *all* applications. (See the International Classification of Goods and Services listed on the inside of the back cover.) At least $200.00 must accompany the application, or the application will be denied a filing date and all the papers returned to the applicant.

Additional Fees Related to Intent-To-Use Applications

In addition to the application filing fee, applicants filing based on having a bona fide intention to use a mark in commerce must submit a fee of **$100.00** for **each class** of goods or services in the application when filing any of the following:

- an AMENDMENT TO ALLEGE USE
- a STATEMENT OF USE
- a REQUEST FOR AN EXTENSION OF TIME TO FILE A STATEMENT OF USE

Miscellaneous Information on Fees

All payments must be made in United States currency, by check, post office money order or certified check. Personal or business checks may be submitted. Make checks and money orders payable to: **The Commissioner of Patents and Trademarks.** The PTO receives no taxpayer funds. The PTO's operations are supported entirely from fees paid by applicants and registrants.

NOTE: FEES ARE NOT REFUNDABLE.

4. SPECIMENS

The following information is designed to provide guidance regarding the specimens required to show use of the mark in commerce.

When an application is based on *use in commerce* the applicant must submit **three specimens** of use of the mark in commerce when filing the original application. If, instead, the application is based on having a *bona fide intention to use mark in commerce*, the applicant must submit **three specimens** at the time the applicant files either an AMENDMENT TO ALLEGE USE or a STATEMENT OF USE.

The specimens must be actual samples of how the mark is actually being used in commerce.

If the mark is used on goods, examples of acceptable specimens are tags or labels which are attached to the goods, containers for the goods, displays associated with the goods, or photographs of the goods showing use of the mark on the goods themselves. If it is impractical to send an actual specimen because of its size, photographs or other acceptable reproductions that show the mark on the goods, or packaging for the goods, must be furnished. Invoices, announcements, order forms, bills of lading, leaflets, brochures, catalogs, publicity releases, letterhead, and business cards generally are not acceptable specimens for goods.

If the mark is used for **services**, examples of acceptable specimens are signs, brochures about the services, advertisements for the services, business cards or stationery showing the mark in connection with the services, or photographs which show the mark as it is used in the selling or advertising of the services. In the case of a service mark, the specimens *must show the mark and include some clear reference to the type of service(s) rendered under the mark.* The three specimens may be identical or they may be examples of three different uses showing the same mark.

Specimens may not be larger than 8½ inches by 11 inches and must be flat. See pages 16 through 18 for samples of some different types of specimens.

INTENT-TO-USE APPLICATIONS - ADDITIONAL REQUIREMENTS

An applicant who files its application based on having *a bona fide intention to use a mark in commerce* must make use of the mark in commerce before the mark can register. After use in commerce begins, the applicant must submit:

1. three specimens evidencing use as discussed above;
2. a fee of **$100.00** per class of goods or services in the application; and
3. either (1) an AMENDMENT TO ALLEGE USE if the application has not yet been approved for publication (use PTO Form 1579) or (2) a STATEMENT OF USE if the mark has been published and the PTO has issued a NOTICE OF ALLOWANCE (use PTO Form 1580).

If the applicant will not make use of the mark in commerce within six months of the NOTICE OF ALLOWANCE, then the applicant must file a REQUEST FOR AN EXTENSION OF TIME TO FILE A STATEMENT OF USE, under 37 CFR 2.89, or the application is abandoned. (Use PTO Form 1581, which is intended only for this purpose.)

See the instructions and information on the back of the forms. The previous information about specimens, identifications of goods and services and dates of use is also relevant to filing an AMENDMENT TO ALLEGE USE or STATEMENT OF USE.

SAMPLE WRITTEN APPLICATION BASED ON USE IN COMMERCE
(Two classes)

TRADEMARK/SERVICE MARK APPLICATION, PRINCIPAL REGISTER, WITH DECLARATION	MARK (Word(s) and/or Design) 1 PINSTRIPES AND DESIGN	CLASS NO. 2 (If known) 16 & 35

TO THE ASSISTANT SECRETARY AND COMMISSIONER OF PATENTS AND TRADEMARKS:

APPLICANT'S NAME: Pinstripes Inc.   3

APPLICANT'S BUSINESS ADDRESS: 100 Main Street   4
(Display address exactly as   Anytown, Missouri 12345
it should appear on registration)

APPLICANT'S ENTITY TYPE: (Check one and supply requested information)

	Individual - Citizen of (Country):	5a
	Partnership - State where organized (Country, if appropriate): _____ Names and Citizenship (Country) of General Partners: _____	5b
X	Corporation - State (Country, if appropriate) of Incorporation: Missouri	5c
	Other (Specify Nature of Entity and Domicile):	5d

GOODS AND/OR SERVICES:

Applicant requests registration of the trademark/service mark shown in the accompanying drawing in the United States Patent and Trademark Office on the Principal Register established by the Act of July 5, 1946 (15 U.S.C. 1051 et. seq., as amended) for the following goods/services (SPECIFIC GOODS AND/OR SERVICES MUST BE INSERTED HERE):
Magazines in the field of business management (Class 16); business management   6
consulting services (Class 35)

BASIS FOR APPLICATION: (Check boxes which apply, but never both the first AND second boxes, and supply requested information related to each box checked.)

XX 7(a)	Applicant is using the mark in commerce on or in connection with the above identified goods/services. (15 U.S.C. 1051(a), as amended.) Three specimens showing the mark as used in commerce are submitted with this application. •Date of first use of the mark in commerce which the U.S. Congress may regulate (for example, interstate or between the U.S. and a foreign country): (Class 16) 1/15/92; (Class 35) 8/27/90 •Specify the type of commerce: Interstate (for example, interstate or between the U.S. and a specified foreign country) •Date of first use anywhere (the same as or before use in commerce date): Cl 16-1/15/92; Cl 35-8/27/90 •Specify manner or mode of use of mark on or in connection with the goods/services: On the magazines and in advertisements for the services (for example, trademark is applied to labels, service mark is used in advertisements)
[ ] 7(b)	Applicant has a bona fide intention to use the mark in commerce on or in connection with the above identified goods/services. (15 U.S.C. 1051(b), as amended.) •Specify intended manner or mode of use of mark on or in connection with the goods/services: _____ (for example, trademark will be applied to labels, service mark will be used in advertisements)
[ ] 7(c)	Applicant has a bona fide intention to use the mark in commerce on or in connection with the above identified goods/services, and asserts a claim of priority based upon a foreign application in accordance with 15 U.S.C. 1126(d), as amended. • Country of foreign filing: _____ • Date of foreign filing: _____
[ ] 7(d)	Applicant has a bona fide intention to use the mark in commerce on or in connection with the above identified goods/services, and, accompanying this application, submits a certification or certified copy of a foreign registration in accordance with 15 U.S.C. 1126(e), as amended. • Country of registration: _____ • Registration number: _____

NOTE: Declaration, on Reverse Side, MUST be Signed

The undersigned, being hereby warned that willful false statements and the like so made are punishable by fine or imprisonment, or both, under Section 1001 of Title 18 of the United States Code and that such willful false statements may jeopardize the validity of the application or any resulting registration, declares that he/she is properly authorized to execute this application on behalf of the applicant; he/she believes the applicant to be the owner of the trademark/service mark sought to be registered, or, if the application is being filed under 15 U.S.C. 1051(b), he/she believes applicant to be entitled to use such mark in commerce; to the best of his/her knowledge and belief no other person, firm, corporation, or association has the right to use the above identified mark in commerce, either in the identical form thereof or in such near resemblance thereto as to be likely, when used on or in connection with the goods/services of such other person, to cause confusion, or to cause mistake, or to deceive; and that all statements made of his/her own knowledge are true and that all statements made on information and belief are believed to be true.

January 16, 1992 _____   [signature]   8
DATE   SIGNATURE

(123) 456-7890 _____   John Doe, Jr., President
TELEPHONE NUMBER   PRINT OR TYPE NAME AND POSITION

SAMPLE WRITTEN APPLICATION
BASED ON INTENT TO USE IN COMMERCE
(One class)

TRADEMARK/SERVICE MARK APPLICATION, PRINCIPAL REGISTER, WITH DECLARATION	MARK (Word(s) and/or Design) 1 THEORYTEC	CLASS NO. 2 (If known) 9

TO THE ASSISTANT SECRETARY AND COMMISSIONER OF PATENTS AND TRADEMARKS:

APPLICANT'S NAME: A-OK Software Development Group   3

APPLICANT'S BUSINESS ADDRESS: 100 Main Street   4
(Display address exactly as   Anytown, Missouri   12345
it should appear on registration)

APPLICANT'S ENTITY TYPE: (Check one and supply requested information)

	Individual - Citizen of (Country):	5a
X	Partnership - State where organized (Country, if appropriate): Missouri Names and Citizenship (Country) of General Partners: Mary Baker, citizen of the USA; Harry Parker, citizen of the USA; and Jane Witlow, citizen of the USA	5b
	Corporation - State (Country, if appropriate) of Incorporation:	5c
	Other (Specify Nature of Entity and Domicile):	5d

GOODS AND/OR SERVICES:

Applicant requests registration of the trademark/service mark shown in the accompanying drawing in the United States Patent and Trademark Office on the Principal Register established by the Act of July 5, 1946 (15 U.S.C. 1051 et. seq., as amended) for the following goods/services (SPECIFIC GOODS AND/OR SERVICES MUST BE INSERTED HERE):
Computer software for analyzing statistics   6

BASIS FOR APPLICATION: (Check boxes which apply, but never both the first AND second boxes, and supply requested information related to each box checked.)

[ ] 7(a)	Applicant is using the mark in commerce on or in connection with the above identified goods/services. (15 U.S.C. 1051(a), as amended.) Three specimens showing the mark as used in commerce are submitted with this application. •Date of first use of the mark in commerce which the U.S. Congress may regulate (for example, interstate or between the U.S. and a foreign country): _____ •Specify the type of commerce: _____ (for example, interstate or between the U.S and a specified foreign country) •Date of first use anywhere (the same as or before use in commerce date): _____ •Specify manner or mode of use of mark on or in connection with the goods/services: _____ (for example, trademark is applied to labels, service mark is used in advertisements)
X 7(b)	Applicant has a bona fide intention to use the mark in commerce on or in connection with the above identified goods/services. (15 U.S.C. 1051(b), as amended.) •Specify intended manner or mode of use of mark on or in connection with the goods/services: On labels affixed to the software (for example, trademark will be applied to labels, service mark will be used in advertisements)
[ ] 7(c)	Applicant has a bona fide intention to use the mark in commerce on or in connection with the above identified goods/services, and asserts a claim of priority based upon a foreign application in accordance with 15 U.S.C. 1126(d), as amended. • Country of foreign filing: _____ • Date of foreign filing: _____
[ ] 7(d)	Applicant has a bona fide intention to use the mark in commerce on or in connection with the above identified goods/services, and, accompanying this application, submits a certification or certified copy of a foreign registration in accordance with 15 U.S.C. 1126(e), as amended. • Country of registration: _____ • Registration number: _____

NOTE: Declaration, on Reverse Side, MUST be Signed

...ents and the like so made are ...001, and that such willful false ... resulting registration, declares ... behalf of the applicant: ...service mark sought to be ...051(b), he/she believes ...est of his/her knowledge and ... right to use the above ...of or in such near resemblance ... goods/services of such other person, to cause confusion, or to cause mistake, or to deceive; and that all statements made of his/her own knowledge are true and that all statements made on information and belief are believed to be true.

February 2, 1992
DATE

Mary Baker   8
SIGNATURE

(123) 456-7890
TELEPHONE NUMBER

Mary Baker, General Partner
PRINT OR TYPE NAME AND POSITION

SAMPLE DRAWING - SPECIAL FORM

8½" (21.6 cm)

APPLICANT'S NAME: Pinstripes Inc.

APPLICANT'S ADDRESS: 100 Main Street, Anytown, MO 12345

GOODS AND SERVICES: Magazines in the field of business management; business management consulting services

FIRST USE: Magazines (Class 16) January 15, 1992
Consulting (Class 35) August 27, 1990

FIRST USE IN COMMERCE: Magazines (Class 16) January 15, 1992
Consulting (Class 35) August 27, 1990

DESIGN: A zebra

Pinstripes

11"
(27.9
cm)

SAMPLE DRAWING - TYPEWRITTEN

8½" (21.6 cm)

11"
(27.9
cm)

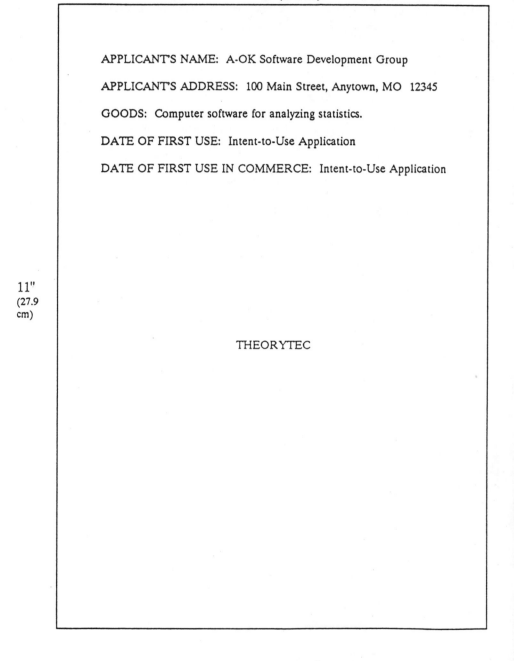

APPLICANT'S NAME: A-OK Software Development Group

APPLICANT'S ADDRESS: 100 Main Street, Anytown, MO 12345

GOODS: Computer software for analyzing statistics.

DATE OF FIRST USE: Intent-to-Use Application

DATE OF FIRST USE IN COMMERCE: Intent-to-Use Application

THEORYTEC

SAMPLE SPECIMEN FOR GOODS (Issue of magazine)

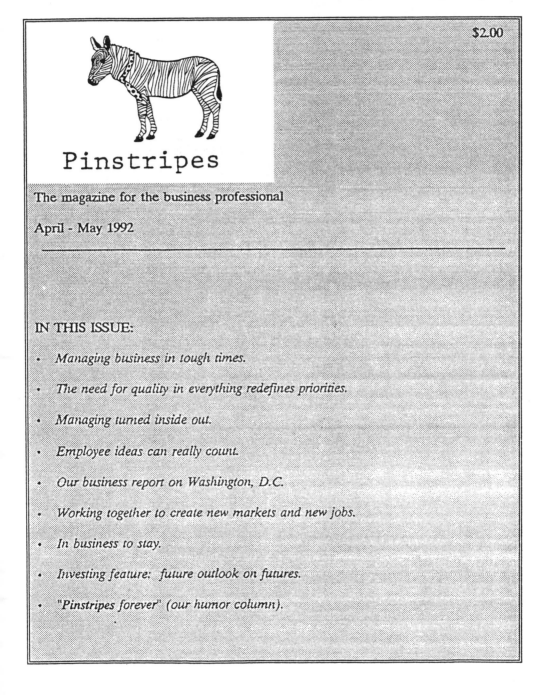

$2.00

Pinstripes

The magazine for the business professional

April - May 1992

IN THIS ISSUE:

- *Managing business in tough times.*

- *The need for quality in everything redefines priorities.*

- *Managing turned inside out.*

- *Employee ideas can really count.*

- *Our business report on Washington, D.C.*

- *Working together to create new markets and new jobs.*

- *In business to stay.*

- *Investing feature: future outlook on futures.*

- *"Pinstripes forever" (our humor column).*

SAMPLE SPECIMEN FOR SERVICES (Advertisement)

Pinstripes

If better business management solutions are what you're after, then think of **Pinstripes** for consulting. We'll come wherever you are to offer a wide range of consulting services for diverse industries, including high-tech fields. You'll like the results, as well as our competitive price.

The more you get to know us, the more you'll realize that we're a best choice for consulting that can make a big difference.

Call or write us.

Pinstripes Inc.
(123) 456-7890 100 Main St., Anytown, MO 12345

SAMPLE SPECIMEN FOR SERVICES
(Business card showing mark <u>and</u> reference to service)

Business Management Consultants

Pinstripes

JOHN DOE, JR
President

100 Main Street, Anytown, MO 12345 USA
(123) 456-7890

SAMPLE SPECIMENS FOR GOODS
(Label affixed to computer disc)

THEORYTEC™

Version 5.0

A-OK Software Development Group

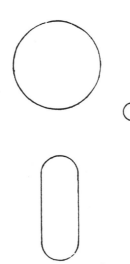

THEORYTEC™

Version 5.0

A-OK Software Development Group

TRADEMARK/SERVICE MARK APPLICATION, PRINCIPAL REGISTER, WITH DECLARATION	MARK (Word(s) and/or Design)	CLASS NO. (If known)

TO THE ASSISTANT SECRETARY AND COMMISSIONER OF PATENTS AND TRADEMARKS:

APPLICANT'S NAME:

APPLICANT'S BUSINESS ADDRESS:
(Display address exactly as
it should appear on registration)

APPLICANT'S ENTITY TYPE: (Check one and supply requested information)

Individual - Citizen of (Country):

Partnership - State where organized (Country, if appropriate): _____
Names and Citizenship (Country) of General Partners: _____

Corporation - State (Country, if appropriate) of Incorporation: _____

Other (Specify Nature of Entity and Domicile): _____

GOODS AND/OR SERVICES:

Applicant requests registration of the trademark/service mark shown in the accompanying drawing in the United States Patent and Trademark Office on the Principal Register established by the Act of July 5, 1946 (15 U.S.C. 1051 et. seq., as amended) for the following goods/services (SPECIFIC GOODS AND/OR SERVICES MUST BE INSERTED HERE):

BASIS FOR APPLICATION: (Check boxes which apply, but never both the first AND second boxes, and supply requested information related to each box checked.)

[ ] Applicant is using the mark in commerce on or in connection with the above identified goods/services. (15 U.S.C. 1051(a), as amended.) Three specimens showing the mark as used in commerce are submitted with this application.
 • Date of first use of the mark in commerce which the U.S. Congress may regulate (for example, interstate or between the U.S. and a foreign country): _____
 • Specify the type of commerce: _____
   (for example, interstate or between the U.S. and a specified foreign country)
 • Date of first use anywhere (the same as or before use in commerce date): _____
 • Specify manner or mode of use of mark on or in connection with the goods/services: _____

 (for example, trademark is applied to labels, service mark is used in advertisements)

[ ] Applicant has a bona fide intention to use the mark in commerce on or in connection with the above identified goods/services. (15 U.S.C. 1051(b), as amended.)
 • Specify intended manner or mode of use of mark on or in connection with the goods/services: _____

 (for example, trademark will be applied to labels, service mark will be used in advertisements)

[ ] Applicant has a bona fide intention to use the mark in commerce on or in connection with the above identified goods/services, and asserts a claim of priority based upon a foreign application in accordance with 15 U.S.C. 1126(d), as amended.
 • Country of foreign filing: _____ • Date of foreign filing: _____

[ ] Applicant has a bona fide intention to use the mark in commerce on or in connection with the above identified goods/services and, accompanying this application, submits a certification or certified copy of a foreign registration in accordance with 15 U.S.C. 1126(e), as amended.
 • Country of registration: _____ • Registration number: _____

NOTE: Declaration, on Reverse Side, MUST be Signed

DECLARATION

The undersigned being hereby warned that willful false statements and the like so made are punishable by fine or imprisonment, or both, under 18 U.S.C. 1001, and that such willful false statements may jeopardize the validity of the application or any resulting registration, declares that he/she is properly authorized to execute this application on behalf of the applicant; he/she believes the applicant to be the owner of the trademark/service mark sought to be registered, or, if the application is being filed under 15 U.S.C. 1051(b), he/she believes applicant to be entitled to use such mark in commerce; to the best of his/her knowledge and belief no other person, firm, corporation, or association has the right to use the above identified mark in commerce, either in the identical form thereof or in such near resemblance thereto as to be likely, when used on or in connection with the goods/services of such other person, to cause confusion, or to cause mistake, or to deceive; and that all statements made of his/her own knowledge are true and that all statements made on information and belief are believed to be true.

DATE

SIGNATURE

TELEPHONE NUMBER

PRINT OR TYPE NAME AND POSITION

INSTRUCTIONS AND INFORMATION FOR APPLICANT

TO RECEIVE A FILING DATE, THE APPLICATION **MUST** BE COMPLETED AND SIGNED BY THE APPLICANT AND SUBMITTED ALONG WITH:

1. The prescribed FEE ($200.00) for each class of goods/services listed in the application;
2. A DRAWING PAGE displaying the mark in conformance with 37 CFR 2.52;
3. If the application is based on use of the mark in commerce, THREE (3) SPECIMENS (evidence) of the mark as used in commerce for each class of goods/services listed in the application. All three specimens may be in the nature of: (a) labels showing the mark which are placed on the goods; (b) photographs of the mark as it appears on the goods, (c) brochures or advertisements showing the mark as used in connection with the services.
4. An APPLICATION WITH DECLARATION (this form) - The application must be signed in order for the application to receive a filing date. Only the following person may sign the declaration, depending on the applicant's legal entity: (a) the individual applicant; (b) an officer of the corporate applicant; (c) one general partner of a partnership applicant; (d) all joint applicants.

SEND APPLICATION FORM, DRAWING PAGE, FEE, AND SPECIMENS (IF APPROPRIATE) TO:

U.S. DEPARTMENT OF COMMERCE
Patent and Trademark Office, Box TRADEMARK
Washington, D.C. 20231

Additional information concerning the requirements for filing an application is available in a booklet entitled **Basic Facts About Trademarks**, which may be obtained by writing to the above address or by calling: (703) 305-HELP.

This form is estimated to take 15 minutes to complete. Time will vary depending upon the needs of the individual case. Any comments on the amount of time you require to complete this form should be sent to the Office of Management and Organization, U.S. Patent and Trademark Office, U.S. Department of Commerce, Washington D.C., 20231, and to the Office of Information and Regulatory Affairs, Office of Management and Budget, Washington, D.C. 20503.

AMENDMENT TO ALLEGE USE UNDER 37 CFR 2.76, WITH DECLARATION	MARK (Identify the mark)
	SERIAL NO

TO THE ASSISTANT SECRETARY AND COMMISSIONER OF PATENTS AND TRADEMARKS:

APPLICANT NAME:

Applicant requests registration of the above-identified trademark/service mark in the United States Patent and Trademark Office on the Principal Register established by the Act of July 5, 1946 (15 U.S.C. 1051 et. seq., as amended). Three specimens showing the mark as used in commerce are submitted with this amendment.

☐ Check here if Request to Divide under 37 CFR 2.87 is being submitted with this amendment.

Applicant is using the mark in commerce on or in connection with the following goods/services:

_____

_____

(NOTE: Goods/services listed above may not be broader than the goods/services identified in the application as filed)

Date of first use of mark anywhere: _____

Date of first use of mark in commerce
which the U.S. Congress may regulate: _____

Specify type of commerce: (e.g., interstate, between the U.S. and a specified foreign country) _____

_____

Specify manner or mode of use of mark on or in connection with the goods/services: (e.g., trademark is applied to labels, service mark is used in advertisements) _____

_____

The undersigned being hereby warned that willful false statements and the like so made are punishable by fine or imprisonment, or both, under 18 U.S.C. 1001, and that such willful false statements may jeopardize the validity of the application or any resulting registration, declares that he/she is properly authorized to execute this Amendment to Allege Use on behalf of the applicant; he/she believes the applicant to be the owner of the trademark/service mark sought to be registered; the trademark/ service mark is now in use in commerce; and all statements made of his/her own knowledge are true and all statements made on information and belief are believed to be true.

_____   _____
Date   Signature

_____   _____
Telephone Number   Print or Type Name and Position

PTO Form 1579 (REV. 9/89)
OMB No 06510023
Exp. 6-30-92

U.S. DEPARTMENT OF COMMERCE/Patent and Trademark Office

INSTRUCTIONS AND INFORMATION FOR APPLICANT

In an application based upon a bona fide intention to use a mark in commerce, applicant must use its mark in commerce before a registration will be issued. After use begins, the applicant must submit, along with evidence of use (specimens) and the prescribed fee(s), either:

(1) an Amendment to Allege Use under 37 CFR 2.76, or
(2) a Statement of Use under 37 CFR 2.88.

The difference between these two filings is the timing of the filing. Applicant may file an Amendment to Allege Use before approval of the mark for publication for opposition in the **Official Gazette**, or, if a final refusal has been issued, prior to the expiration of the six month response period. Otherwise, applicant must file a Statement of Use after the Office issues a Notice of Allowance. The Notice of Allowance will issue after the opposition period is completed if no successful opposition is filed. Neither Amendment to Allege Use or Statement of Use papers will be accepted by the Office during the period of time between approval of the mark for publication for opposition in the **Official Gazette** and the issuance of the Notice of Allowance.

Applicant may call (703) 557-5249 to determine whether the mark has been approved for publication for opposition in the **Official Gazette.**

Before filing an Amendment to Allege Use or a Statement of Use, applicant must use the mark in commerce on or in connection with all of the goods/services for which applicant will seek registration, **unless** applicant submits with the papers, a request to divide out from the application the goods or services to which the Amendment to Allege Use or Statement of Use pertains. (See: 37 CFR 2.87, Dividing an application)

Applicant **must** submit with an Amendment to Allege Use or a Statement of Use:

(1) the appropriate fee of $100 per class of goods/services listed in the Amendment to Allege Use or the Statement of Use, and

(2) three (3) specimens or facsimiles of the mark as used in commerce for each class of goods/services asserted (e.g., photograph of mark as it appears on goods, label containing mark which is placed on goods, or brochure or advertisement showing mark as used in connection with services).

Cautions/Notes concerning completion of this Amendment to Allege Use form:

(1) The goods/services identified in the Amendment to Allege Use must be within the scope of the goods/services identified in the application as filed. Applicant may delete goods/services. Deleted goods/services may not be reinstated in the application at a later time.

(2) Applicant may list dates of use for only one item in each class of goods/services identified in the Amendment to Allege Use. However, applicant must have used the mark in commerce on all the goods/services in the class. Applicant must identify the particular item to which the dates apply.

(3) Only the following person may sign the verification of the Amendment to Allege Use, depending on the applicant's legal entity: (a) the individual applicant; (b) an officer of corporate applicant; (c) one general partner of partnership applicant; (d) all joint applicants.

This form is estimated to take 15 minutes to complete. Time will vary depending upon the needs of the individual case. Any comments on the amount of time you require to complete this form should be sent to the Office of Management and Organization, U.S. Patent and Trademark Office, U.S. Department of Commerce, Washington D.C., 20231, and to the Office of Information and Regulatory Affairs, Office of Management and Budget, Washington, D.C. 20503.

| STATEMENT OF USE UNDER 37 CFR 2.88, WITH DECLARATION | MARK (Identify the mark) |
| | SERIAL NO. |

TO THE ASSISTANT SECRETARY AND COMMISSIONER OF PATENTS AND TRADEMARKS:

APPLICANT-NAME:

NOTICE OF ALLOWANCE ISSUE DATE:

Applicant requests registration of the above-identified trademark/service mark in the United States Patent and Trademark Office on the Principal Register established by the Act of July 5, 1946 (15 U.S.C. 1051 et. seq., as amended). Three (3) specimens showing the mark as used in commerce are submitted with this statement.

☐ Check here only if a Request to Divide under 37 CFR 2.87 is being submitted with this Statement.

Applicant is using the mark in commerce on or in connection with the following goods/services: (Check One)

☐ Those goods/services identified in the Notice of Allowance in this application.

☐ Those goods/services identified in the Notice of Allowance in this application except: (Identify goods/services to be deleted from application) _____

_____

_____

Date of first use of mark anywhere: _____

Date of first use of mark in commerce which the U.S. Congress may regulate: _____

Specify type of commerce: (e.g., interstate, between the U.S. and a specified foreign country) _____

_____

Specify manner or mode of use of mark on or in connection with the goods/services: (e.g., trademark is applied to labels, service mark is used in advertisements) _____

_____

The undersigned being hereby warned that willful false statements and the like so made are punishable by fine or imprisonment, or both, under 18 U.S.C. 1001, and that such willful false statements may jeopardize the validity of the application or any resulting registration, declares that he/she is properly authorized to execute this Statement of Use on behalf of the applicant; he/she believes the applicant to be the owner of the trademark/service mark sought to be registered; the trademark/ service mark is now in use in commerce; and all statements made of his/her own knowledge are true and all statements made on information and belief are believed to be true.

Date _____

Signature _____

Telephone Number _____

Print or Type Name and Position _____

INSTRUCTIONS AND INFORMATION FOR APPLICANT

In an application based upon a bona fide intention to use a mark in commerce, applicant must use its mark in commerce before a registration will be issued. After use begins, the applicant must submit, along with evidence of use (specimens) and the prescribed fee(s), either:

> (1) an Amendment to Allege Use under 37 CFR 2.76, or
> (2) a Statement of Use under 37 CFR 2.88.

The difference between these two filings is the timing of the filing. Applicant may file an Amendment to Allege Use before approval of the mark for publication for opposition in the **Official Gazette**, or, if a final refusal has been issued, prior to the expiration of the six month response period. Otherwise, applicant must file a Statement of Use after the Office issues a Notice of Allowance. The Notice of Allowance will issue after the opposition period is completed if no successful opposition is filed. Neither Amendment to Allege Use or Statement of Use papers will be accepted by the Office during the period of time between approval of the mark for publication for opposition in the **Official Gazette** and the issuance of the Notice of Allowance.

Applicant may call (703) 557-5249 to determine whether the mark has been approved for publication for opposition in the **Official Gazette**.

Before filing an Amendment to Allege Use or a Statement of Use, applicant must use the mark in commerce on or in connection with all of the goods/services for which applicant will seek registration, unless applicant submits with the papers, a request to divide out from the application the goods or services to which the Amendment to Allege Use or Statement of Use pertains. (See: 37 CFR 2.87, Dividing an application)

Applicant must submit with an Amendment to Allege Use or a Statement of Use:

> (1) the appropriate fee of $100 per class of goods/services listed in the Amendment to Allege Use or the Statement of Use, and

> (2) three (3) specimens or facsimiles of the mark as used in commerce for each class of goods/services asserted (e.g., photograph of mark as it appears on goods, label containing mark which is placed on goods, or brochure or advertisement showing mark as used in connection with services).

Cautions/Notes concerning completion of this Statement of Use form:

> (1) The goods/services identified in the Statement of Use must be identical to the goods/services identified in the Notice of Allowance. Applicant may delete goods/services. Deleted goods/services may not be reinstated in the application at a later time.

> (2) Applicant may list dates of use for only one item in each class of goods/services identified in the Statement of Use. However, applicant must have used the mark in commerce on all the goods/services in the class. Applicant must identify the particular item to which the dates apply.

> (3) Only the following person may sign the verification of the Statement of Use, depending on the applicant's legal entity: (a) the individual applicant; (b) an officer of corporate applicant; (c) one general partner of partnership applicant; (d) all joint applicants.

This form is estimated to take 15 minutes to complete. Time will vary depending upon the needs of the individual case. Any comments on the amount of time you require to complete this form should be sent to the Office of Management and Organization, U.S. Patent and Trademark Office, U.S. Department of Commerce, Washington D.C., 20231, and to the Office of Information and Regulatory Affairs, Office of Management and Budget, Washington, D.C. 20503.

REQUEST FOR EXTENSION OF TIME UNDER 37 CFR 2.89 TO FILE A STATEMENT OF USE, WITH DECLARATION	MARK (Identify the mark)
	SERIAL NO.

TO THE ASSISTANT SECRETARY AND COMMISSIONER OF PATENTS AND TRADEMARKS:

APPLICANT NAME:

NOTICE OF ALLOWANCE MAILING DATE:

Applicant requests a six-month extension of time to file the Statement of Use under 37 CFR 2.88 in this application.

☐   Check here if a Request to Divide under 37 CFR 2.87 is being submitted with this request.

Applicant has a continued bona fide intention to use the mark in commerce in connection with the following goods/services: (Check one below)

☐   Those goods/services identified in the Notice of Allowance in this application.

☐   Those goods/services identified in the Notice of Allowance in this application except: (Identify goods/services to be deleted from application) _____

_____

This is the _____ request for an Extension of Time following mailing of the Notice of Allowance.
(Specify first - fifth)
If this is not the first request for an Extension of Time, check one box below. If the first box is checked, explain the circumstance(s) of the non-use in the space provided:

☐   Applicant has not used the mark in commerce yet on all goods/services specified in the Notice of Allowance; however, applicant has made the following ongoing efforts to use the mark in commerce on or in connection with each of the goods/services specified above:

_____

If additional space is needed, please attach a separate sheet to this form

☐   Applicant believes that it has made valid use of the mark in commerce, as evidenced by the Statement of Use submitted with this request; however, if the Statement of Use is found by the Patent and Trademark Office to be fatally defective, applicant will need additional time in which to file a new statement.

The undersigned being hereby warned that willful false statements and the like so made are punishable by fine or imprisonment, or both, under 18 U.S.C. 1001, and that such willful false statements may jeopardize the validity of the application or any resulting registration, declares that he/she is properly authorized to execute this Request for Extension of Time to File a Statement of Use on behalf of the applicant; he/she believes the applicant to be the owner of the trademark/service mark sought to be registered; and all statements made of his/her own knowledge are true and all statements made on information and belief are believed to be true.

_____   _____
Date   Signature

_____   _____
Telephone Number   Print or Type Name and Position

PTO Form 1581 (REV. 9/89)
OMB No. 06510023
Exp. 6-30-92
                                     U.S. DEPARTMENT OF COMMERCE/Patent and Trademark Office

INSTRUCTIONS AND INFORMATION FOR APPLICANT

Applicant must file a Statement of Use within six months after the mailing of the Notice of Allowance in an application based upon a bona fide intention to use a mark in commerce, UNLESS, within that same period, applicant submits a request for a six-month extension of time to file the Statement of Use. The request must:

(1) be in writing,

(2) include applicant's verified statement of continued bona fide intention to use the mark in commerce,

(3) specify the goods/services to which the request pertains as they are identified in the Notice of Allowance, and

(4) include a fee of $100 for each class of goods/services.

Applicant may request four further six-month extensions of time. No extension may extend beyond 36 months from the issue date of the Notice of Allowance. Each request must be filed within the previously granted six-month extension period and must include, in addition to the above requirements, a showing of GOOD CAUSE. This good cause showing must include:

(1) applicant's statement that the mark has not been used in commerce yet on all the goods or services specified in the Notice of Allowance with which applicant has a continued bona fide intention to use the mark in commerce, and

(2) applicant's statement of ongoing efforts to make such use, which may include the following: (a) product or service research or development, (b) market research, (c) promotional activities, (d) steps to acquire distributors, (e) steps to obtain required governmental approval, or (f) similar specified activity .

Applicant may submit one additional six-month extension request during the existing period in which applicant files the Statement of Use, unless the granting of this request would extend beyond 36 months from the issue date of the Notice of Allowance. As a showing of good cause, applicant should state its belief that applicant has made valid use of the mark in commerce, as evidenced by the submitted Statement of Use, but that if the Statement is found by the PTO to be defective, applicant will need additional time in which to file a new statement of use.

Only the following person may sign the verification of the Request for Extentsion of Time, depending on the applicant's legal entity: (a) the individual applicant; (b) an officer of corporate applicant; (c) one general partner of partnership applicant; (d) all joint applicants.

This form is estimated to take 15 minutes to complete. Time will vary depending upon the needs of the individual case. Any comments on the amount of time you require to complete this form should be sent to the Office of Management and Organization, U.S. Patent and Trademark Office, U.S. Department of Commerce, Washington D.C., 20231, and to the Office of Information and Regulatory Affairs, Office of Management and Budget, Washington, D.C. 20503.

Foreign Applicants and General Information

Domestic Representative

Applicants not living in the United States must designate in writing the name and address of some person resident in the United States on whom may be served notices of process in proceedings affecting the trademark. This person will also receive all official communications *unless* the applicant is represented by an attorney in the United States.

Communications With the Patent and Trademark Office

The application and all other communications should be addressed to:

The Commissioner of Patents and Trademarks
Washington, DC 20231

It is preferred that the applicant indicate his or her telephone number on the application form. Once a serial number is assigned to the application, the applicant should refer to this number in all telephone and written communications concerning the application.

Additional Information

The federal registration of trademarks is governed by the Trademark Act of 1946, 15 U.S.C., Sec. 1051 et seq.; the Rules, 37 C.F.R., Part 2; and the Trademark Manual of Examining Procedure.

Information Phone Numbers

General Trademark or Patent Information: (703) 557-INFO

Status Information for Particular Trademark Applications: (703) 557-5249

General Copyright Information: (202) 479-0700

Trademark Appendix III

The International Schedule of Classes of Goods and Services

Goods

1. Chemical products used in industry, science, photography, agriculture, horticulture, and forestry; artificial and synthetic resins; plastics in the form of powders, liquids, or pastes, for industrial use; manures (natural and artificial); fire-extinguishing compositions; tempering substances and chemical preparations for soldering; chemical substances for preserving foodstuffs; tanning substances; adhesive substances used in industry

2. Paints, varnishes, and lacquers; preservatives against rust and against deterioration of wood; coloring matters and dyestuffs; mordants and natural resins; metals in foil and powder form for painters and decorators

3. Bleaching preparations and other substances for laundry use; cleaning, polishing, scouring, and abrasive preparations; soaps; perfumery, essential oils, cosmetics, and hair lotions; dentifrices

4. Industrial oils and greases (other than oils and fats and essential oils); lubricants; dust-laying and -absorbing compositions; fuels (including motor spirit) and illuminants; candles, tapers, night lights, and wicks

5. Pharmaceutical, veterinary, and sanitary substances; infants' and invalids' foods; plasters and material for bandaging; material for filling teeth, dental wax, and disinfectants; preparations for killing weeds and destroying vermin

6. Unwrought and partly wrought common metals and

their alloys; anchors, anvils, bells, rolled and cast building materials; rails and other metallic materials for railway tracks; chains (except driving chains for vehicles); cables and wires (nonelectric); locksmiths' work; metallic pipes and tubes; safes and cash boxes; steel balls; horseshoes; nails and screws; other goods in nonprecious metal not included in other classes; ores

7. Machines and machine tools; motors (except for land vehicles); machine couplings and belting (except for land vehicles); large-size agricultural implements; incubators

8. Hand tools and instruments; cutlery, forks, and spoons; side arms

9. Scientific, nautical, surveying, and electrical apparatus and instruments (including wireless); photographic, cinematographic, optical, weighing, measuring, signaling, checking (supervision), life-saving, and teaching apparatus and instruments; coin or counter-feed apparatus; talking machines; cash registers; calculating machines; fire-extinguishing apparatus

10. Surgical, medical, dental, and veterinary instruments and apparatus (including artificial limbs, eyes, and teeth)

11. Installations for lighting, heating, steam-generating, cooking, refrigerating, drying, ventilating, water-supply, and sanitary purposes

12. Vehicles; apparatus for locomotion by land, air, or water

13. Firearms; ammunition and projectiles; explosive substances; fireworks

14. Precious metals and their alloys, and goods in precious metals or coated therewith (except cutlery, forks, and spoons); jewelry and precious stones; horological and other chronometric instruments

15. Musical instruments (other than talking machines and wireless apparatus)

16. Paper and paper articles, and cardboard and cardboard articles; printed matter, newspapers, periodicals, and books; bookbinding material; photographs; stationery and adhesive materials; artists' materials; paint brushes;

typewriters and office requisites (other than furniture); instructional and teaching material (other than apparatus); playing cards; printers' type and cliché (stereotype)

17. Gutta-percha, india rubber, balata and substitutes, and articles made from these substances and not included in other classes; plastics in the form of sheets, blocks, and rods, being for use in manufacture; materials for packing, stopping, or insulating; asbestos, mica, and their products; hose pipes (nonmetallic)

18. Leather and imitations of leather, and articles made from these materials and not included in other classes; skins and hides; trunks and traveling bags; umbrellas, parasols, and walking sticks; whips, harnesses, and saddlery

19. Building materials, natural and artificial stone, cement, lime, mortar, plaster, and gravel; pipes of earthenware or cement; roadmaking materials; asphalt, pitch, and bitumen; portable buildings; stone monuments; chimney pots

20. Furniture, mirrors, and picture frames; articles (not included in other classes) of wood, cork, reeds, cane, wicker, horn, bone, ivory, whalebone, shell, amber, mother-of-pearl, meerschaum, and celluloid, and substitutes for all these materials, or of plastics

21. Small domestic utensils and containers (*not* of precious metals, *or* coated therewith); combs and sponges; brushes (other than paint brushes); brushmaking materials; instruments and material for cleaning purposes and steel wool; unworked or semiworked glass (excluding glass used in building); glassware, porcelain, and earthenware, not included in other classes

22. Ropes, string, nets, tents, awnings, tarpaulins, sails, and sacks; padding and stuffing materials (hair, kapok, feathers, seaweed, etc.); raw fibrous textile materials

23. Yarns and threads

24. Tissues (piece goods); bed and table covers; textile articles not included in other classes

25. Clothing, including boots, shoes, and slippers

26. Lace and embroidery, ribbons, and braid; buttons, press buttons, hooks and eyes, and pins and needles; artificial flowers
27. Carpets, rugs, mats, and matting; linoleums and other materials for covering existing floors; wall hangings (nontextile)
28. Games and playthings; gymnastic and sporting articles (except clothing); ornaments and decorations for Christmas trees
29. Meats, fish, poultry, and game; meat extracts; preserved, dried, and cooked fruits and vegetables; jellies and jams; eggs, milk, and other dairy products; edible oils and fats; preserves and pickles
30. Coffee, tea, cocoa, sugar, rice, tapioca, sago, and coffee substitutes; flour, and preparations made from cereals; bread, biscuits, cakes, pastry, and confectionery; ices; honey and treacle; yeast and baking powder; salt, mustard, pepper, vinegar, sauces, and spices; ice
31. Agricultural, horticultural, and forestry products and grains not included in other classes; living animals; fresh fruits and vegetables; seeds; live plants and flowers; foodstuffs for animals; malt
32. Beer, ale, and porter; mineral and aerated waters and other nonalcoholic drinks; syrups and other preparations for making beverages
33. Wines, spirits, and liqueurs
34. Tobacco, raw or manufactured; smokers' articles; matches

Services

35. Advertising and business
36. Insurance and financial
37. Construction and repair
38. Communication
39. Transportation and storage
40. Material treatment
41. Education and entertainment
42. Miscellaneous

Index

Advertising, 59-60
Agents
  licensing, 47-49, 94-98
  literary, 141-143
  patent, 14-16, 68
Attorneys, patent, 14-16, 57, 68, 83

Basic Facts About Trademarks, 68
Bern Convention for the Protection
  of Literary and Artistic Works,
  185-186
Better Business Bureau, 48-49
Businesses, investigating, 51-52

Character names, trademarking,
  193
Classification Definitions, 67-68
"Conception of the invention," 27
Copyrights, 5-6, 111-112. *See also*
    Creative works, publication of
  applicants, requirements for, 134
  application forms, 153-184
  and collaborations, 118-119, 137
  for collections, 132-133, 161-168
  corrections to, 133, 177-180
  eligibility of works for, 111,
    114-117, 118
  and "fair use," 120, 121-123
  fees, 130, 135, 156
  future trends and, 151-152

and government works, 117,
  126-127
history of, 112-113
infringement upon, 149-151
international, 185-187
limits on, 120-121
notice, use of, 124-128
and public domain material, 117
registering, 129-132, 133, 134-136
renewals of, 137-138, 149, 181-184
rights granted by, 113-114,
  137-138
searches, of records, 136
and works for hire, 119-120, 144,
  161-168
Creative works, publication of,
  139-141
  agents and, 141-143
  licensing agreements, 144-149
  presentation package, 143-144

*Directory of Registered Patent Attorneys
  and Agents Arranged by States
  and Countries*, 68
Disclosure documents, 16-17

"Fair use," 120, 121-123

Goods and services, classifications
  of, 209-210, 231-234

Hodgson, Paul, 65
Hopkins, Samuel, 20

Ideas, 3-4
    evaluating, 11-13
    as intellectual property, 4-5
    marketing. See Creative works,
        publication of; Inventions,
        marketing
    protecting. See Copyrights;
        Patents; Trademarks
    showing to businesses, 13-14, 16
Index of Patents, 67
Index of Trademarks, 67
Inventions, marketing
    advertising, 59-60
    agents, licensing, 47-49, 94-98
    and "invention companies," 46
    licensing agreements, 49-51, 62,
        99-104
    manufacturing, 53-56
    raising money, 56-58
    selling the patent, 52-53, 60-62

Lawsuits
    and copyrights, 149-151
    and patents, 62-64
    and trademarks, 207
Letter of intent, 57
Liability, 50-51
Libraries, patent and trademark,
    72-75, 205-206
Library of Congress Catalog
    Numbers, 131, 188
Licensing agreements
    and copyrights, 144-149
    and patents, 49-51, 62, 99-104
Literary Market Place, 142

Manual of Classification, 67
Manual of Patent Examining
    Procedure, 68
Manufacturing, of inventions, 53-56

1976 Copyright Act, 113, 115, 117,
    120, 127, 137, 148, 156
Nondisclosure agreements, 7, 14,
    49, 105-107

Official Gazette, 66-67

Paris Convention for the Protection
    of Industrial Property, 81-82,
    84
Patent and Trademark Office
    corresponding with, 69-70
    history of, 20-23, 68
    publications of, 66-68
Patent Cooperation Treaty, 82
Patents, 5. See also Inventions,
    marketing
    applicants, requirements for,
        26-27
    application forms, 90-93
    applications, contents of, 28-38
    attorneys/agents for, 14-16, 57, 68,
        83
    claims, 18, 32-33, 63
    corrections to, 40-41
    design patents, 41-42
    and disclosure documents, 16-17
    drawings and prototypes for, 12,
        32, 33-37
    eligibility of ideas for, 23-26
    fees, 16, 38-40, 76-80
    infringement upon, 62-64
    international, 81-85
    national defense exception, 39
    and nondisclosure agreements, 7,
        14, 49, 105-107
    patent-pending status, 45
    plant patents, 42-44
    publications about, 66-68
    rights granted by, 18-20
    sample, 86-89
    search libraries, 71-75
    selling, 52-53, 60-62
    and trademarks, 192-193

Plant Variety Protection Act, 44
Press releases, 59
Prior art, 25-26
Public domain material, 117

"Reduction to practice," 27

Service marks, 191
Silly Putty, 64-65
Statute of Anne, 112
*Story of the United States Patent Office,
    The,* 68
Supplemental Register, 205

Title 37 Code of Federal Regulations,
    68
Trade secrets, 6-8
Trademark Official Gazette, 195
Trademarks, 5, 191
    applications, contents of, 197-203
    application forms, 208-229
    and character names, 193
    drawings for, 200-202, 211-212,
        217

fees, 203, 212-213
goods and services, classifications
    of, 209-210, 231-234
grounds for refusal of, 204-205
infringement upon, 207
international, 230
notice, use of, 206
and patented products, 192-193
publications about, 66-68
registering, 193-196, 203-204
rights granted by, 191-192
sample, 215-221
search libraries, 205-206
specimens for, 202-203, 213,
    219-221

United States Constitution, 112-113
Universal Copyright Convention
    (UCC), 185

Winslow, Samuel, 20
Works for hire, 7-8, 119-120, 144,
    161-168
Wright, James, 64-65

About the Creative Group

The Creative Group was founded in 1991 to provide assistance to inventors, artists, writers, and entrepreneurs in the licensing of intellectual property rights. Its president, Gary Ahlert, is a successful inventor, writer, and entrepreneur who has had professional experience with each step outlined in this book. The organization provides a full range of services, including product consultation and development, patent searches and applications, trademark searches and applications, copyrights, licensing, and marketing.

The Creative Group
400 Main Street
Stamford, CT 06901
(203) 359-3500